The Winds of Time

The Winds of Time

Sy Varlen

VANTAGE PRESS
New York

FIRST EDITION

Published by Vantage Press, Inc.
516 West 34th Street, New York, New York 10001

Manufactured in the United States of America
ISBN: 0-533-10464-5

Library of Congress Catalog Card No.: 92-96924

0 9 8 7 6 5 4 3 2 1

At the vertex of the eons,
when whole eternities elapsed
and new eternities began,
the winds of time
brought on men to be.

Thereupon,
a new natural force, in full command
of its destiny, would be determinant
in the events to develop
on earth and beyond.

The winds of time
turned momentous
as men set off to advance.

—Sy Varlen

Contents

The Winds of Time

I. – An Eternal Start

The speed of light is not constant.
If it were, light would be eternal. Not being eternal, light should eventually disappear, suddenly and completely.

But eternity and disappearance
are both unaccountable, for light is not eternal, nor its speed constant.

Relatively few rays of starlight
ever strike inert targets. Then, some rays are absorbed. Others are reflected. Most rays never hit anything inert.

On their nimble departure from stars,
light rays travel for eons through the void of space. Through the eons, they lose momentum. Light rays and all radiation slow down at a constant rate.

Space grows larger for older radiation
and will suggest an expanding universe if a constant speed of light is the only means of measuring the cosmic vastness.

Critical speed lost,
ultimately dispersed, exhausted light corpuscles cease to be radiant. A time comes when corpuscles take a last undulatory beat. Their intrinsical electromagnetic nature persists, however.

Positive, negative,
pervading space; each positive bit appeals the opposite. The call felt, they engage in rotary motion, one around the other.

Undulatory motion of rays
having stopped, rotary motion of atom components begins instead. Space curves for new substance and new gravitation.

Whereas radiant energy
proceeds instantly from matter in intense action, matter develops slowly from the debris of radiation.

If light were eternal
or if it disappeared completely, there would be no regeneration of the universe and hydrogen clouds constantly forming throughout space would be unaccountable as to origin and ends.

The first and only atoms
to assemble from formerly radiant corpuscles are hydrogen atoms. Hydrogen, the simplest, the lightest, the most active and most abundant, is the element of elements.

The universe is a hydrogen universe.
Nebulas are immense clouds of hydrogen atoms in process of condensation into stellar masses. New stars then submit to the gravitational order of galaxies. On growing vastly larger and apart, nebulas may give rise to new galaxies.

Stars shine
while their hydrogen plasma fuses into helium. Other elements float in space as dust clouds and stay inert unless hydrogen power gives them a shape and a destiny.

There is no degeneration of energy
toward an eventual universal warmth. Heat does not warm up space. Heat only excites the molecules of substance. Out of substance, in the void of space, there is nothing to excite.

Thence, heat does not dissipate into space,
for it travels as fast as light, as infrared radiation, without warming up the vacuum.

A warm universe is not physically possible.
All along its existence, matter may undergo any degree of temperature, density or rarity, but no inert stage will be final—absolute cold being the basic temperature in the depths of space.

Space is timeless.
Matter and radiation, by their dynamics and duration, are the factors of time. Atoms and rays can last for billionths of a second or for billions of years. They mark the passing of time, which is the rhythm of change.

Nothing is permanent but change.
Vast amounts of matter and radiation, locked up and annihilated in alleged black pits, would in time explode as superconcentrated gravity.

Such fantastic explosions
would create new space, plus boundless quantities of matter and radiation, all of a sudden, in the guise of quasars and bursting galaxies. Local bing bangs then would be the only accountable bing bangs—if ever.

II. – Spheres and Orbits

Stars are round.
Roundness is induced by newly formed forces of gravitation acting concentrically within fluid concentrations of substance.

Larger gaseous concentrations
develop stronger gravitational attraction; hydrogen atoms then reach higher degrees of excitement, and the new stars will grow bigger and hotter than the average luminous bodies.

If stars are born too big,
growing too hot and bright, they will explode as novas too soon. Then they become white dwarfs, and their superconcentrated mass exerts unusually strong gravitation.

Middle-sized stars,
like the earth's sun, shine through the billions of their planets' yearly revolutions.

Also round,
because of their own gravitation, planets attract cosmic dust, bits and chunks of stray matter, in an endless process of accretion.

Planets revolving close to a star
receive more concentrated radiation. They become solid and smaller, for they lose their lighter elements sooner.

When inner planets rotate slowly,
their sunlit sides heat up all the more, and their dark sides cool down sharply, notwithstanding the proximity of the source of radiation. If enveloped in heavy atmospheres, slow-spinning planets may grow entirely hot and dim.

Planets in outer orbits
receive rarified radiation. In the nearly interstellar cold, outer planets keep their primary atmosphere and enormous sizes. Some can even form rings of frozen gas.

Planetary orbits are not closed circuits,
being rather spirals traced as planets circle traveling stars. Ignoring their tiny followers, stars travel in vastly wider orbits around remote centers of galaxial gravitation.

Whereas planets escort stars
from whose very nebular substance were formed, satellites are minor bodies, generally captured by planets.

But if far enough from strong planetary attractions,
small bodies trace independent orbits round a star,
comets and asteroids being their possible varieties.

Bright spheres illuminate the universe.
Dark spheres, most of them forever lifeless, gyrate by gravitational balance, thereby prevented from plunging into the stars and from escaping into space.

III. – Life Inevitable

Desolation under the suns
is a monotonous theme throughout the universe.
Cratered rocky worlds and cold fluid globes constitute
the ignored retinues of most stars.

Too distant to have joined the mass of a star,
even in the nebular stage, too close to escape its pull,
too small to turn hot enough to shine, planets are
leftovers from the processes of star formation.

A sun with no planetary retinue
would be as improbable as a roundly shaped nebula
condensing into a sphere, without leaving any spare
material at all.

Still, the mass of planets,
as compared with that of stars, is so negligible that,
with no planets anywhere, the cosmic balance of matter
and radiant energy would be about the same and stars
would shine the same as well.

The abysmal chill
of the void surrounds stars all over. Extreme cold
prevailing over warmth assigns most lands below the
suns a universal fate of desolation.

Yet, for all the universal harshness,
the living state is a potentiality of active substance. There is a recondite tendency in matter to submit its elements to the order of life.

Life is the highest triumph of matter,
a cosmic trend to an ideal state. The power of life is the power of active substance to reach a higher status over inert matter.

Life is the power of energy
seeking subtler outlets through the minds of higher beings. Life is a chain reaction, like blazing fire and hydrogen fusion.

There is no leap from dust to flesh
by magical spells or alien agents. Life emerges freshly, naturally, gradually, inexorably by irreversible upward steps wherever the vitally active presence of surface liquid water and sunlight last enough to sustain biological progress.

Planets, though, are barren by nature.
Deadly hostile worlds, most of them cold, normally lack surface flowing water. Many never harden into solid spheres.

On the sunny earth, water had to flow plentifully.
The dry plains of the young planet were to be neither dry nor plain forever. Newly formed, the earth was hot and violent.

Volcanoes emerged as the first mountains on earth, spewing lava, ashes, steam. Lava would flow briefly to cool off into stone. Ashes precipitated. But steam, resulting from oxydized rocks and hydrogen under great pressure and heat, invaded the raw methane-ammonia atmosphere.

Growing steam clouds
condensed into water droplets. It started to rain. Yet, on a planet that had been dry and hot through the times following its igneous stage of formation, rains were all but heavy or regular. Pools and ponds gathered only to evaporate and disappear.

As the accumulation of steam clouds increased,
rains grew heavier. Cracks, old impact craters and lowlands were inundated. Warm shallow seas formed everywhere, and the expansion of the waters would proceed for the time to come.

The increasingly thick clouds
generated electrical charges. Intense lightning action incorporated nitrogen in the medium. Great vital developments were then to occur.

All over, photosynthetic processes
broke water into free oxygen and hydrogen. Minerals and carbon oxydes readily combined with the oxygen being freed, giving off sugars. Sugars built up colloids, which would acquire high complexity as albumins and proteins.

Colloidal concentrations
suspended in the vital medium of quiet waters then lodged complex proteinic motile reticula vibrating by induction of electrical discharges from the atmosphere.

Agitating within the protective,
nourishing colloid-albumin walls, lively motile reticula rose into existence as the protoplasm of living cells.

The complexity and richness
of the aquatic medium moved toward its culminating synthesis by promoting inorganic active substance to the organic order of the living cell.

IV. – The Dare of Life

Life in a world of water was perfectly natural.
A miracle would have been a lifeless world full with water. A still greater miracle would have been life prospering in a dry world.

Water flowed on earth
for reasons as compelling as those forbidding its circulation in other worlds.

Powerfully compelling reasons
for the steadfast presence of water on earth were a moderate distance from the sun, a speed of rotation determining a superb thermal balance, a planetary mass exerting enough gravity to retain air and water vapor—the elements of a permanent pluvial system.

On becoming a mighty terrestrial element,
water caused the medium to be hospitable and prompted life to arise. Once life did, it had the way to progress wide open. Rather, life had no choice but to advance.

Progress required time,
and time there was. A lasting sun gave terrestrial life time to advance boundlessly. Had the sun lasted shorter, liquid water and any trace of life on earth would have disappeared long ago.

The immense task of making organic
the inorganic primitive seas took the first billion years of the biological history of the earth.

In amply varied forms of algae,
starting from the microscopic, plant cells, the first living forms ever, proved productive rather than consumptive. The increasing production of sugars, carbohydrates, and protein made urgent the necessity of a new system of life to consume the plentiful vegetal food.

So it happened
that some of the most advanced types of plant cells turned to modify their structures and their functions. Instead of breaking carbon dioxide into carbohydrates and free oxygen, as plants did, they would breathe the already abundant oxygen to burn carbohydrates. This was an opposite process developed by the very plants to restore biological balance.

Such parasitic system of life
was to grow into the animal kingdom. Through the next billion years, the Archeozoic seas became saturated with plant cells supporting an enormous variety of restless animal cells.

When the two billion years
of the Archeozoic era ended, and the Proterozoic began, animal cells were associating to originate multiple classes of minute organisms.

Just emerging from the vegetable world,
the first animals were very simple and small. They barely amounted to tiny stomachs catching still tinier bits of food suspended in water. Many species belonged to the group of polyps.

Associating in large colonies,
some species of polyps built extensive coral reefs, in an impressive manifestation of the power of life, even at its emerging stage, to mark the physical medium significantly, no matter how small each individual could be.

Sponges, too, spread out
to populate the marine beds. Attached to the sea bottom, like the polyps, sponges were lowly animals drawing water through their pores to extract organic nutrients.

Sponges were so simple
and the sea water so rich in organic matter that sponges never needed to evolve. In their elementary nature, sponges inhabited the first seas of the earth, and were to be living in the last.

The immobile abundance of sponges
in the late Pre-Cambrian seas heralded the emergence of more complex and active animals through Paleozoic times.

The large family of mollusks
was first in growing muscles to work articulate organs. Oysters could open and close their shells, snails crawled slowly along, and cuttlers swam by expelling water backward.

Another great group of invertebrates
was formed by the arthropods, which included the active crustaceans. More active than the echinoderms—widely represented by starfishes—or than most mollusks, trilobites were the earliest and most common of crustaceans.

The segmented bodies and jointed limbs
of Crustacea gave crabs enough mobility to swim and also to walk briefly by the shores of the barren land. Insects were to emerge later as the first animals to make to the air in wings, projecting the triumph of Arthropoda far ahead.

The great explosion of invertebrate life
having occurred during the Cambrian was caused by the diffuse superabundance of organic matter, accumulated by primitive life forms up to the end of the Proterozoic era.

The abundance of tiny organisms consumed,
larger creatures had to eat smaller ones, hence becoming separated into vegetarians and carnivores. Biological balance was again restored, and widely extended species of food particle eaters, like the trilobites, perished.

The world of invertebrates
was a world of quiet waters. But as the Paleozoic advanced, the aquatic medium turned more active. It rained more heavily. Great rivers formed and opened courses over arid lands that were not so plain as before.

Such a fluvial system gave the earth's surface
a liquid dynamics no sister planet would ever have. The seas grew in extension and depth. Marine currents and streams gradually flowed through the seas and regulated the world climate by blending tropical and polar waters.

The liquid pressure of the seas
fractured the old crust into continental plates. These plates were bound to shift constantly over the asthenosphere, and were pushed and drawn by the unpredictable fluid forces of the oceanic mass, so creating the peculiarly terrestrial phenomenon of plate tectonics. Trenches and chasms opened in the porous hot lithosphere.

Mountains were pushed up to restore balance.
The deeper the trenches opened along continental shelves the higher maritime mountain ranges rose, and the more seismic coastal lands became over the subduction zones.

Vast original flat lands
split into abrupt continents. The irruption of great rivers of fresh water altered the cozy marine medium.

The water pouring from the lands
was lighter, sweet, richly oxygenated. Simple creatures as the invertebrates were, especially those encased in shells and crusts, would not feel the difference very acutely.

But a few,
endowed with keener senses throughout their soft bodies, relished the impact and acquired the power to swim upstream.

It was some unknown kind
of slender sensitive chordates that by undulating exertion countered the downward rush of incoming rivers. In the effort they strengthened the rudimentary muscles along the chords.

Later, by the reward in excitement and success
inducing to more effort, the chords themselves hardened into articulate pieces of harder tissue. Spines were under way, thus giving even primitive fishes the mastery of rivers and seas.

Indeed, the main motive effort of fishes
was undulation by elongated bodies tapering off backward. Fins later developed as accessory means of propulsion, steering, and equilibrium. Thus, a fish, like no invertebrate whatsoever, was a bundle of muscles framed by a spine and governed by a heading brain through the nerves converging to the spinal cord.

The whole fish body
was streamlined, as it was primarily intended to swim in swift water. And when rivers ran too shallow, some fishes would creep, aided by their fins.

With nothing but air to breathe,
gills would have to work as lungs. With no food particles suspended in running water, the first jawed fishes appeared, chewing solid bits of food and nipping on plants.

Toothed jaws subsequently developed
into countless forms of tearing and triturating batteries, arming aggressive vertebrate carnivores with the deadliest weapons in the animal world.

By enabling themselves
to live outside water, especially adventurous types of fish evolved into amphibians. However, living only part time on the land, without completely leaving the water, kept some genera of amphibians as a lowly transitional class.

Yet, most fish species
kept their fish status to prosper as masters of the aquatic world, with all the anatomical advances made by swimming upstream. Higher types of Amphibia eventually advanced to originate the wholly terrestrial class of reptiles.

The beginning of the Mesozoic era
marked the time when reptiles invaded the land at large. By then, algae had already developed terrestrial plant forms and undertaken the first green revolution over the brown land. Up to that time, the terrestrial surface not covered by water had been utterly lifeless.

Having experimented
with countless rootless, leafless types of plants like the psilophytes, then filling swamps and shallows with a large variety of aquatic herbs, the vegetable kingdom advanced to cover the land with extensive fern forests.

That was the world that reptiles enjoyed,
and where they advanced and dominated with no opposition and no competition for over a hundred million years of the Mesozoic era.

The unrestrained growth of land vertebrates,
manifest in the size and variety of dinosaurs, was nurtured by the green revolution that had been going on since the Silurian, without animal consumers.

That amazing growth of vertebrates
was a land version of the invertebrate explosion in the Paleozoic seas, at its time brought about by unicellular and other tiny organisms pervading the seas since Pre-Cambrian times.

The long era of dinosaurs
definitely stated the excellence of backbones. Terrestrial vertebrates acquired the power to creep, walk, run, and fly. Those returning to the sea resumed the fish shape, although breathing air with terrestrial lungs.

The aquatic habitat kept luring
with the comfort of a constant temperature. Safe from abrupt changes, aquatic life was as safe from the urge to change. Simple species undergoing little or no change would survive age after age.

In the strange world of the lands,
the living found that only change was permanent. Survival from one day to the next was uncertain. Midday heat gave way to the chill of the night.

Swamps would dry up
to become wastelands. At one time or another, prairies would turn into deserts, and the lowlands and shallows of one epoch might form the highlands of another.

Yet, the raw open space was stimulating.
The transparent air expanded the vision and eased all efforts, but survival would take a superb ability to evolve. Fast mutation into multiple new forms or plain extinction were the alternatives of life ashore.

Dinosaurs opened the lands
to the emergence of higher types of vertebrates. By developing colossal bodies, they perfected the mechanics of efficient skeletons.

By perishing to the last of their genera,
dinosaurs submitted to the fact that the terrestrial medium changed faster than they could. As the earth's axis tilted more pronouncedly, dragged by the fluid inertia of the expanding seas, summers and winters succeeded perpetual springs.

The great swamps and fern forests
extant up to the Jurassic vanished. Deserts, conifer forests, and steeper mountains featured much of the Cretaceous scenery.

Being cold-blooded,
dinosaurs could not stand the new climatic extremes for very long. Besides heat and cold, voracious saurian carnivores evolved in later times and primitive mammals preying on eggs reduced the numbers of vegetarian reptiles, which never meant overpopulation given their mostly huge sizes.

Moreover, as reptiles had not been
notorious migrants, they were ill-equipped for long walks. When the time came to migrate in search of the last good lands, most could not go far. As diplodocus died out, tyrannosaurus followed suit.

Seen in succession,
trilobites had abounded in all the seas and dinosaurs grew to enormous sizes as land animals owing to extraordinary biological circumstances of food plenty. When those circumstances—although long-lasting—were altered, trilobites and then dinosaurs were up for extinction.

The sea allowed sharks
to voracious survival, whereas the lands provided dinosaurs with plenty of dust to lie under. Still, primitive sharks in very many varieties had been plundering the seas a hundred million years before dinosaurs walked the lands.

The abrupt changes
closing the Cretaceous further prompted dinosaurs toward the realm of the extinct. By then, a new class of vertebrates was well on its way to take over. To them, change bore a challenge and a chance, not merely to survive, but to win by fast adaptation.

The novel masters
were not coming from a different order. They had sprung from the very reptile design, though bringing along vital changes and improvements.

While reptiles laid eggs
and left them at the mercy of the sun and predators, the new class would bear the unborn young until delivering them alive.

Another gigantic step
in the way to success was taken. The new class grew able to suckle their newborn until they could chew solid food.

A third major stride forward
was warm blood. Up to the Cretaceous, the animal kingdom had been thoroughly cold-blooded. Warm blood, which reptiles never developed, brought about crucial advantages.

The revolutionary change
introduced by warm blood in the metabolic system was aided by the growth of hair. Hair helped keep constant body temperatures for uninterrupted action both in hot and cold weather, when reptiles would remain lethargic.

The new system was expensive.
Keeping the body warm when all else was cold required more food. This economic problem, though, worked to the advantage of the hairy innovators. A higher requirement of food made them more active and aggressive. They were the mammals.

As the Cenozoic began,
mammals already stood as the highest zoological class and had the earth as a legacy. The mighty saurians had departed forever. The terrestrial landscape was then animated by a large variety of mammalian species. They were fit to thrive in nearly every climate possible under the sun.

Insects had for long been flying through the air; ungainly reptiles had been crawling, walking, flying, swimming, but it was the superbly restless mammalian class that gave the world an image of movement.

The great migrations
in an endless search for wider horizons soon made mammals universal. Brains grew more complex to process the data streaming in through ever keener senses.

Most species of mammals, however,
did not go much further in brain development. They specialized in food and habits from the beginning. Anteaters would be eating ants, and moles digging burrows down to the end.

But brains would continue to improve
in those making good use of them. Intelligent species would not solve their survival problems by growing armors or by dulling brains in the comfort of social security.

As the Cenozoic era advanced,
the earth changed even faster. More water vapor forming under the crust filtered out into the air. Heavy rainfall inundated more lowlands and raised sea levels.

The ever increasing volumes of water
flowing on the planet caused the fragments of old Pangaea to drift away as widely separated continents and archipelagos.

How far continental masses
drifted from the equator toward the poles determined their climate. The character of plants and animals was a result of local conditions.

In warm areas
the variety of living forms grew amazingly. By far, Africa became the most prolific of the continents.

South America, by contrast,
with an often flooded Amazon basin, a barren expanse to the northeast, and the steep Andean mountain range rising along the whole west coast, failed to shelter steady zoological progress and also to build up a substantial organic soil, excepting that of the southeastern Pampas.

A separate land mass,
Mesozoic North America obviously enjoyed a warm climate and was home to the greatest success of dinosaurs on earth. Yet, the North American mammals of the Tertiary, even evolving fast, had less time to reach advanced forms.

Glaciations struck the North
and curtailed evolutionary advances. Primitive primates appeared only briefly in the North, which suggested that higher primates like men would have never emerged in any of the two American continents.

Without the haven
of a stably hospitable continent, animate terrestrial life could not have surpassed the mammalian beast stage, not even over the whole Quaternary period.

It was Mother Africa,
an immense plateau safe from sea invasions and great floods, that throughout the Tertiary provided mammals with a unique chance to reach their finest forms. Besides, Africa was far from the chill of glaciations preventing the progress of life in higher latitudes during the late Pliocene.

It was in Africa, and nowhere else,
that men of various types emerged, rather compulsorily, from the ranks of higher primates then being in possession of exclusive evolutionary advances. With no specialized habits or features at all, they surpassed the highest levels of animaldom to reach those of reasoning.

Lacking paws, they developed opposing thumbs
to climb to trees for protection, as all primates had done. Opposing thumbs then helped the ablest of them, who belonged to the Homo genus, wield sticks for weapons.

Sticks, bones, stones
were soon being used also as tools. An erect position was permanently adopted as the forelimbs were spared locomotive work.

Well before the end of the Pliocene,
the first types of men existed. They were good walkers and also craftsmen with increasingly able hands. They reasoned with brains that improved as fast as the needs of the race multiplied.

Their keen sense of cooperation
called for a means of communication by articulate sounds, later arranged as languages.

Oral communication
greatly aided to teach the young and organize vastly powerful collective action. With those exclusive assets and an unquenchable thirst for adventure, men set out to march northward. It was the dawn of the Pleistocene. Africa was turning sultry, but was still connected to Europe.

V. - Of Glories and Downfalls

Association and cooperation
set men on the road to supremacy. When many individuals put up thoughts and efforts on a definite purpose, no obstacle could hinder their progress.

But association by self-interest
soon became dependence. By using their wits better, a few rose to dominate, and the rest submitted. Since they were first formed, human groups agreed on living under someone's authority. Honorary at first, government posts before long became the best paying jobs everywhere.

As agriculture turned hunters and gatherers
into farmers, binding them to the land, societies grew larger and permanent. The use of metals led to refinement in all orders of life. Inevitably, gold whetted the appetite for riches.

The human person was degraded to slavery,
as unpaid labor came to be a part of the social system. Empires stretched out from societies largely submitted to the will of men wanting ever more.

Great glory and wealth
many enjoyed indeed, though the victors and plunderers of one round would be the beaten and the sacked of the next.

Egypt began the writing of history
with a supreme chapter. There was only one Nile River
and one feat like the Egyptian challenge to time.

Founded by the end of the Near Eastern Neolithic,
when the desertification of North Africa had reached
the Nile, and after Sumer had laid the basis of city life,
the Egyptian was the most enduring of civilizations.

Drawing a line where the uncertainties
of prehistory ended and a consistent history began, the
dynasties of Pharaohs spanned all the time taken by
successive Middle Eastern empires to rise and vanish.
The Tigris and the Euphrates were certainly not the
Nile.

A most elongated oasis,
Egypt for long throve, protected by the deserts, not by
mountains or fortifications. In a land of blue skies,
Egyptians enjoyed the dry air of the deserts and the
rich waters from the tropics farther south.

Most enjoyable was the land
where floods did not devastate, but yearly deposited the
riches of fertility. The Nile, however, had to be tamed by
steady hard work that required cooperation and stable
government. It was the most benign god river that men
ever worshiped.

A world in itself,
Egypt cared little about projecting her culture beyond the Nile valley. Conquest and exploration were not national objectives, for the cult of death and the preparation for the other life were.

In the way to splendor
Egypt had no deadly enemies to overcome. So, it was not a nation of soldiers or captives. Its incredible monuments showed no signs of systematic destruction.

Having no angry gods to placate, Egyptians had no taste for bloody sacrifices. Most truly, when passing to the realm of Osiris, kings and subjects claimed by writing to have been good and just.

Men and women of superb good taste,
Egyptians dressed their slim frames barely when humble; transparently when proud. Women were perfect in the art of fascination. Visibly, the charm of God Nile's daughters was a joyous motive of Egyptian art.

The god-kings' obsession
of raising colossal structures expressed their wealth and power as it meant durable pain to the laborers. With nothing better than bronze tools and muscle traction, cyclopean labor on rationed food was permanent through the dynasties of builder Pharaohs.

By the time Cleopatra's tragedy
closed the history of the Egyptian New Empire, the great Mesopotamian kings belonged to the past. Babylon's restless annals of
splendor-destruction-revival had ended with the Persian conquest. Mostly built with sun-baked bricks, Babylon left no wall standing from what had been magnificent palaces, towers, fortresses.

Assyrian Nineveh was razed to the ground
by those who so hated it. Assyrians at best left memories of military efficiency and bestial cruelty. Amid the rubble they also left large amounts of fascinating clay tablets and bas-reliefs.

Then, the short dynasty of Persian kings
had time for little but extensive conquest on the generalized decadence of West Asian kingdoms. An effective system of communications was created to articulate the first driving empire in the world. The Persian civilization lacked the time to produce a seasoned culture. Prematurely corrupt, Persia fell in days to the mercy of Alexander of Macedon.

Alexander stormed the East
in his obsession to emulate the far-reaching Persian drives. He then built an empire that could have hardly outlived him, even if he had lived longer. Boundless ambition plus heavy drinking and a fever marked him as a cataclysmic conqueror and innovator rather than the founder of a lasting empire.

Although of Eastern stock,
individualist Hebrews heralded the coming of the Western era long before the Macedonians of Alexander. Israel did little to convert the non-Jewish, since Judaism was racist by nature, but its Messianic Scriptures motivated a large number of conflicting interpretations over the West.

Following Alexander's irruption into the East,
the center of world power began to shift west, to Europe, and west it kept shifting. The fall of Persia and that of Carthage had in fact opened the Western era.

The Athenian culture was to stand
as the fountain of inspiration to the rising Western world. Not a state of kings and subjects, but one of free citizens, Athens was a school of art with a vision of eternity; a discipline of thinking for the introspection of subjective quest and the adventure of objective learning.

Having originated
a distinctly Western conception of the fine arts and the first method of scientific thought, the Grecian masters inaugurated the Western rank of the creative genius.

In times when Eastern societies
marked the passing of the years by the succession of kings, Athenians brightened their course with a constellation of artists, philosophers, statesmen.

Without the light of Athens
the Western culture would not have been Western.
Short of better inspiration, Europe would have turned
East for culture as it did for religion. Without the Greek
spirit, which was common to Rome, neither Darius nor
Hannibal would have been beaten.

The direct heirs and greatest propagators
of Greek culture were the Romans. Latinized Hellenism
identified the new nucleus of power, clearly intent on
advancing with only the world's size for a limit.

VI. – The Free and the Bond

Very few species in the animal kingdom
had a taste for liberty. Very many would struggle from birth to death just to keep alive, while others, like sea lilies, plainly vegetated. Even fewer had a taste for fun.

Condors, the master fliers,
glided over the roof of the world, free and dominant, although in endless search for food, as nestlings would not wait. Dolphins truly relished life, frisking on the waves, freer than the breeze.

In their feline demeanor,
cats roamed the world uninhibited. Free from work, free from obligation, free from love, family, and society, cats ranked far above chain-loving dogs.

Older in the story of life,
suspended by vibrating wings that developed before vertebrates existed, flies flew for fun as well as for business, greatly enjoying their short days of mischief.

Except in childhood,
men had little time for frolicking. As their numbers augmented, their liberties dwindled. They gladly exchanged the joys of living in the open for the confinement in crowded towns.

With no natural organs for defense or attack, primitive naked humans were very humble. Even with crude weapons, men had to associate to hunt, while flesh-eaters on four made the kill mostly alone. Making war turned men into soldiers, discipline being more important than their own lives.

As farmers, men fell into thorough dependence. Since the first farm was toiled, men became subject to the whims of the weather, to say nothing of rulers, tax collectors, parasites, invaders. People had to pay for protection as they did for submission.

Human liberties were more a notion than a fact. Meanwhile, tribes and clans grew into city-states, then into kingdoms and empires. Stronger nations dominated and exploited simple peoples. Paradoxically, individuals in large countries were more subject to the state than those living in small communities.

VII. – On Being and Having

The function of being
was conditioned to the right to have. Within advanced societies, needs were more complex in order to thrive at ever rising standards of living.

Early men needed little more than food,
but in their virgin world, food was scarce. Those having something to eat every day would have been happier than people of wealth in later times.

From food and shelter,
the range of human possessions and expectations widened as the times improved. Every advance, in any field of knowledge, would improve the status of humans as a whole, for culture was at length to be one over the world.

But as men were not equal
nor had they equal opportunities, the benefits of progress would not reach all of them. There would always be the active and the passive, the winner and the loser.

Progress was a mixed blessing,
dispensing benefits and miseries as it proceeded. More benefits for the few at the top; miseries, of course, for the many at the bottom.

One of the dear products of civilization
was the stratification of social classes, which could be better understood as economic levels.

Nobody with good economic means
would remain in the lower classes without an impulse to climb up the social ladder. It was also impossible to keep a high social status with little or no money.

The political sciences
in charge of the distribution of wealth always failed in their assignment to make the poor less poor. It was far easier to make the rich poor, as bad fortune and social revolutions amply demonstrated.

The philanthropic endeavor
of alleviating the pains of the poor went ineffectual wherever poverty was endemic. Social justice was a slogan of politicians to win adepts, but rarely a practical objective, the results being conspicuous in every large city in the world.

Recurrently high rates of unemployment
kept millions of able people languishing, in a tragic waste of human energy. Then the sick, the old, the illiterate, the unskilled, would find life even harder.

In the form of improved medical care,
material progress contributed to unemployment by giving way to overpopulation. Large cities attracted rural migration, depopulated agricultural areas, and grew beyond the worst expectations.

The problem was indeed complex,
for one thing was to give occupation to all those able to work, and another to provide for the millions unable to help themselves.

Still, there was one way
to promote social improvement in spite of difficulties existing or to exist. It was the truly efficient democratic system of government, which was conceived to keep civil liberties effective in order to forestall tyranny and injustice, the breeders of poverty.

If democracy,
as the only sensible system of government, could not banish poverty, it had to be because it was not real democracy. It might rather be a disguised tyranny of officials helping themselves instead of helping the people.

In a true democracy,
the right to work for a decent pay, to own property, and to prosper through lawful private activities had to make people free from want. That basic freedom was for people the reason to live, and for nations the soundest source of strength.

Prosperous people made prosperous nations
with the power to undertake effective programs to cure the sick, comfort the old, educate the young, and train the unskilled. Poverty, illiteracy, and unemployment then had no reason to exist.

But the moment a democratic government
chose to run big business enterprises, it would be practicing capitalism of state, which constituted the economic system advocated by planners at war with practical sense.

Capitalism of state
was a direct way to economic disaster, naturally provoked by mismanagement and corruption, the built-in evils of state enterprises.

Private capitalism itself
was an economic system by no means directed to social ends. It was efficient business based on the principle of profits. Without efficiency and profits, capitalism would not have had that name.

Nor was true democracy a welfare system.
The government of the free was a clear set of political procedures, whose survival and success depended on keeping liberties effective and favoring a climate of business confidence at all times.

When more independent professionals,
craftsmen, tradesmen, farmers, and corporate concerns prospered in a country, more people were employed and more revenue filled the treasury. Then, nations grew rich and strong if they were honestly administered.

True democracy
was the solution to the abuses of power by absolute monarchies and self-appointed autocracies, all lacking systems of selective succession to power.

As soon as civilization had evolved enough,
Greece and then Rome emerged as the first democracies and the greatest civilizations of their time.

Endless improvement of legal procedures
established democracy as the perfectible political system for all the times, preventing the electoral mechanics in many countries from being a business venture for wealthy candidates, providing also that executives and legislators must be practical, honest, and daring.

VIII. – The Grace of Hers

An exquisite point in biological progress
was reached when beautiful women took shape.
Feminine features were refined to agree with every
advancing standard of evolutionary beauty.

The differences between the human female
and her conflicting mate were deeper than those
existing between sexes in other species of higher
mammals. Of delicate bodies and smaller brains,
women instinctively turned disadvantages into assets.

What a man perceived
as he met an attractive woman was that he was facing
one of the most fascinating creatures living on earth.

The power of women
sprang from no physical or mental force, but from
almighty charm. Charm was the power no other kind of
creatures had wielded until ladies adopted it as their
ultimate weapon.

Right from the cradle,
the sweetness of little girls was the sum of all the grace
in the perceptible world. While the human race kept
advancing along evolutionary routes, feminine beauty
would be growing subtler and keener.

The simple imagination
of those assigning bestial looks to extraterrestrial humans in popular movies suggested how far they were from surmising the strange, incredible beauty of women at far higher stages of evolution.

Many could deform human faces;
none could conceive them more beautiful than real. Yet, maids were to look prettier in each passing generation. They were to be more intelligent as well.

The progress of societies
led females to study and think more than their mothers and grandmothers ever did.

Feminine creative geniuses, however,
would not occur normally. More by chance than by design, a few women would succeed in government, as this was an enlarged form of housekeeping.

Poetesses, novelists, painters,
arising with frequency and success ever since the fair sex started feeling free, were proof to the new stand of women as men's intellectual peers.

Still, the supreme creativity
of women was revealed by conceiving sons of genius and sweet, charming daughters.

In retrospect,
women were treated by men rather diversely. More unfairly than fairly, mainly owing to the relative feminine corporeal frailty.

For long epochs,
according to race, country, and the old institution of marriage, women were men's possessions. Infidelity deserved physical punishment when not death, as it could not have been otherwise in a world of male chauvinists.

Unquestionably,
feminine behavior toward a man was revealing of his personality. A woman loving her mate faithfully was to behave differently from one who was disappointed in hers, or scared of his jealousy. Not all men were lovable, or fine gentlemen.

Females were customarily spared
high learning and military duties to be prepared for wedlock; to start with, for housekeeping, both concerns being monotonous and likely to dull every brain luster.

As to whether ladies
were going to be as free as men in social and sexual matters, it was evident that females were acquiring more independence as societies became more liberal.

Even so, females had an innate tendency
to be committed to the sterner sex, largely due to physical limitations related to their reproductive functions.

Her inborn sense of reliance
would thus place a woman closer to a man, which made the world a beautiful place to abide. Self-renewing, self-perfecting, women's eternal magic was the greatest inspiration for all the arts and all the dreams.

IX. – The Lands of Men

At first, the lands of the earth
were only one land. There were no continents, for there were no oceans. The earth was a hot flat desert where mountains had yet to come up.

A few billion years later,
water covered the earth, but some solid surface still emerged. Sea bottoms, drifting continents, and islands remained from what had been a wholly dry surface.

Deep trenches
containing water at high pressure prevented the lands still dry from being also submerged.

If no dry land had been left,
earthly living forms would have been solely aquatic. Only algae and invertebrates would have thriven in the shallow areas of a wholly liquid medium, solid only at the frozen polar caps.

The shape and position
of continental plates depended upon watery forces subject themselves to centrifugal forces of the planetary rotation, as well as to the thermal effects of solar radiation and internal heat. Solar and lunar gravitational pulls combined for the complexity of ocean dynamics.

It was despite and because
of strong opposing forces that men emerged as another planetary force. Men were a conflicting power themselves; the best friends, and the worst foes to their own kind.

Although air was most vital to men,
no record of quarreling over air was ever entered. There was air enough for all and no one needed to breathe more air than necessary, so air was neither hoarded nor taken away from others.

Water came next as vitally needed.
As it was still abundant, people were seldom deprived of water save in drought, warfare, torture, or accident.

As the third essential,
food proved very effective to expose miserable traits of the human soul. The rich had food in excess. The poor were starved to the point of undernourishment.

As acquired needs,
gold and money motivated greed. Gold rushes mobilized folks more readily than fanaticism or patriotic causes.

While the meek
expected happiness somewhere beyond life, gold holders could buy the like right on earth. By the possession of wealth, golden dreams would be closer to coming true.

But it was land, the vital space,
that turned men into murderous beasts. Large-scale massacres were committed by invaders, exterminating populations of seized territories.

A group of people without land
would be only a band of nomads or a lot of slaves. To form a nation, a group had to own land. And then, overcrowded countries grew explosive. Their threat against thinly peopled neighbors sooner or later spelled penetration and usurpation.

Politics was an alternate form of war,
causing far less bloodshed, yet the fallen were often prominent contenders. History was strewn with executions and assassinations of kings, liberators, statesmen, tyrants.

Since man was a political animal,
politics was everyone's concern, and the struggle for power was intensely personal as it could be remotely impersonal.

The absolute power of a despot,
on his overthrow any day, might not be captured by his special enemies, but fall into the hands of obscure opportunists, or dissolve into anarchy.

Pieces of the lands of men
were tokens of war, causing suffering for a time, but political power was the object of perpetual conspiracy.

The great adventure
of conquering the world began when men had become good walkers and reasoned that the lands lying north could be worth the risk of leaving home not to return.

Thereupon,
walking in search of paradise beyond Africa, men invaded all the lands; even the Arctic Circle, by the power of adaptation of their Eskimo representatives.

Nomadic groups
would behave distinctly from each other, owing to precarious living conditions. But it was the land of long abode that defined the character and physical features of the scattered groups later called races.

Families settling the tropics
were tinted dark by vertical sun rays and made lethargic by year-long summers. Turned away from temperate areas by those having settled first, other tribes had to migrate farther north and became blond in the boreal twilight. Long winters left them no choice but hard work.

Persistent glacial winds
from Siberia, in northeast Asia, and solar glare from ice and snow made people narrow their eyes into slits, thereby conforming the Mongolian type of people settling over East Asia.

Neolithic Siberian Mongolians
migrated en masse farther east from the northeastern end of Asia, along a bridge set up by glaciations. They reached a promised land on a virgin continent. A score of millennia elapsed since, without fading the strong Mongolian looks of most aboriginal Americans.

Still farther east, in southern Europe,
societies of handsome people were moved to great advancement and refinement by the gentle breezes of the Mediterranean.

The Mediterranean peoples of Europe
assimilated the progress mankind had made since the earliest Mesopotamian-Egyptian and East Indian achievements, and created a distinct Western culture. Without the Eastern heritage, Europeans would have taken much longer to develop a purely European culture.

From her strategic geographical position
on the western end of the Mediterranean, at the gateway of the Atlantic, Spain traced the map and expanded the horizons of the post-Renaissance Western world.

When the question of the flatness
or the roundness of the earth was a direct concern of nearly no one else in Europe, Spain sent Columbus and Magellan on royal missions of discovery.

Where the ocean ended to the west,
Columbus did not find the end of the world— or Asia. He found virgin lands that were wider and richer than the whole of Europe.

Nothing before,
not even the Crusades, or the Polos' account of the wonders of old China, had so dramatically broadened the interests of Western men.

On the first voyage round the world in history, Magellan's westward expedition finally pushed men's minds past the notion of a flat earth.

Since the fall of Rome,
a millennium had passed before a European nation stood up as a world power again.

Spain did it.
Fresh from winning her own Crusades against the Moors, Spain conquered vast American native empires and built an enlarged semblance of Europe and of herself.

The lands that for three centuries
had formed the Spanish overseas provinces, and those of the English, French, Dutch, and Russian colonies in the New World, eventually constituted a colossal bulwark of Western civilization.

Eighteen republics speaking Spanish
plus millions of Hispanics hardly assimilating Anglo-Saxon ways in the United States resulted from the thorough Hispanization of nearly a half of the New World.

Spain was intensely European.
Having been the staunchest Roman province, it became the melting pot of Western Europe. Spain was also African by the Moorish blood that fertilized her soil for seven centuries.

From the legions of explorers,
conquerors, administrators, priests, soldiers of fortune coming home one day with something of America in their blood, Spain became in a measure American, too.

The Roman genius
was not scientific or intellectual. It was rather political and military. The progress of science under the Romans was practically null despite Pliny's solitary investigations and Vitruvius's ignored inventiveness.

The world did not go mechanical
until the English put steam to work machinery and locomotives. Elsewhere in Europe few people seemed to have foreseen how important machines were to become.

By running the first big factories
with steam power, water boilers bloodlessly brought about the Industrial Revolution, though smearing life with the black of soot.

Out of the national character,
as much as through the ethics of Protestantism, Britons inaugurated a new approach to riches by way of large-scale industrial production. This made the West grow practical, busier, wealthier.

Darwinian disclosures
on biological evolution by natural selection reinstated scientific thought two millennia after the Athenians had been opening the way to reason. The immutability of species then had to enter the file of the old misconceptions in biology.

Newton's universe obeying the laws
of gravitation inspired Einstein to lift physics into the heights of relativity. Ten scores of years after Newton, the Atomic Age was already at sight.

The British projection overseas
indeed went as far as the feats of resident islanders. Across the ocean, self-committed English colonists founded an American continental power that would reflect Britain's image through time and space.

Revealing of the British nature
was the fact that history recorded nothing about geniuses of the kind of Buonaparte or Hitler springing from British soil. Far from that, Britons played the firemen's part in times of Continental conflagration.

The model of a sound democracy
was also British. Bravely, though peacefully, the English asserted their liberties as early as the thirteenth century.

When King John had to sign the Magna Charta,
not only the English set out to be free. All those who could feel the difference between liberty and tyranny learned that liberty was possible after all.

And, of course,
the speech of William Shakespeare and Samuel Johnson was garlanded with preference everywhere. Most practical, most beautiful in sound and letter, for all its inconsistencies, English ascended cum laude to the rank of a universal language.

In all countries
more people turned to learn the utterance of English words for pleasant talk and vital communication.

The British contribution
to a better life was beyond appraisal. Likewise, it was hard to imagine when the Industrial Revolution would have taken place without the English practicality.

The part of France
in European history was definitely central. After Hannibal's motley armies, those rallied in Gaul by Vercingetorix were the biggest Rome had to fight on European battlefields.

The conquest of Gaul
meant to Rome the mastery of Western Europe. It also marked the inevitable transition from republican to imperial times.

Centuries later,
on the Roman decline, it was from the vast Gallic country that the Goths brought hell to the lands of Greece and Italy.

The French role in Europe
was even more central when Charlemagne made of his Frankish kingdom the successor of Imperial Rome. His Holy Roman Empire was born with the Middle Ages and died with the French Revolution.

Also a French conception,
the Encyclopédie was a way to systematic learning that led to the age of Enlightenment.

What the Western world
owed France was still more than a taste for culture, science, or liberty. For precise wording and refined expression, French was used as the international language by kings, diplomats, and scholars, succeeding Latin and preceding English.

The German intellectual influence
was intense since the sixteenth century. Luther initiating the Protestant Reformation, Beethoven taking music into intellectualism, Goethe exalting the universality of art and culture, Nietzsche proclaiming the greatness of the human spirit, Einstein placing nuclear power in men's hands, or Marx pronouncing profits evil, each exposed the nature of the German genius.

The German states came to be determinant
in the European balance of power since they were unified. Yet, seventy-four years after unification, Germany was divided again, as it tried to unify Europe by force.

The continuation of the Great War
and a consequence of the sanctions imposed on Germany at Versailles in 1919, World War II at terrible cost in lives and wealth brought on a new era in the history of civilization.

The Germany emerging from so much suffering
had a new vision of the future. In a few decades after nearly total destruction, the nation prospered amazingly and contributed substantially for programs of international cooperation.

As the long cold war came to a close,
the German reunification was to affect Europe as much as other crucial developments did through the dramatic German history.

Second only to Greece
in importance for the emergence of Western civilization, and a most captivating country, Italy presented three aspects of intense interest.

The first aspect
was the natural beauty of the Italian peninsula, caressed by the Adriatic and the Tyrrhenian arms of the Mediterranean. The unique attraction of the aquatic ways of Venice, the warm Italian charm of Naples, the unbelievable beauty of the Isle of Capri showed Italy as the ever young piece of earth that gave Europe its air of a dreamland.

The second Italian marvelous facet
was the suggestion of the remains of classic Rome. The Forum, the Colosseum, the unearthed city of Pompeii, the still fresh beauty of Hadrian's Tivoli transported the mind to the Roman world at its greatest splendor.

The third and culminating aspect of Italy
was her world of Renaissance art. Incredibly but truly, there were the creations of Michelangelo as real objects of contemplation. The supreme marble sculptures of Pietà and David, the paintings of the Sistine Chapel ceilings were all existing together with countless other masterpieces executed by the constellation of Italian Renaissance geniuses.

The importance of Switzerland
kept no proportion with her small geographical extension. Switzerland was a federation of practical people who made the best of their beautiful land and whose strength came from their indeclinable pacifism and neutrality. The high quality of the Swiss industrial production was a token of sustained prosperity.

The Rhine River
had its source among the Swiss Alps, and the country presented an invigorating landscape of snow-capped mountains, swift currents, and blue lakes. Land travel was thus rather slow, so the Swiss skies were crowded with a dense air traffic.

The Swiss were efficient people.
With no seaports or navigable rivers, the Swiss overseas commerce and export trade were more intense than those of many countries exposed to seas and oceans.

Politically,
Switzerland was the first working confederation in Europe, and a model for peoples seeking unification, even when there were old ethnic, religious, and linguistic differences.

The steadfast Swiss neutrality
was vital to the nation's independence; it was also vital to the Central European military balance, as it kept the strategic Alpine heights and passes in reliable neutral hands.

Russia was a colossus
with a national identity still in the making. China had had a long-lived civilization by the time frigid Russia was not even populated.

Acquiring a religion
as late as the tenth century, falling to the Tartar yoke in the thirteenth, shifting to Czarist autocracy on the Tartars' decline, Russia was made home to advanced ideological doctrines without having reached a degree of cultural development comparable with that of the rest of Europe.

The conglomerate of Caucasians,
Tartars, Mongols, and Muslims was unified in the rough fashion the populations of all the Russias so well tolerated.

Further unifying factors
were the invasions of great European armies in 1812 and 1941, which, far from making of Russia a land of conquest, rather consolidated the nation for all her big size and the ethnic diversity of the population.

Devoted to constitute
a bicontinental power since Peter the Great created St. Petersburg as a great city and the Russian north European stronghold, Russia, as the twentieth century USSR, became capable of deciding much of the unfolding of world events—for as long as it could keep the strength of a superpower.

Europe had been the vital space for Roman expansion for three centuries. Then, on the fall of Rome, Europe lay helpless. Without the Byzantine Empire stopping the Turks, Europe would have fallen to them and become Muslim long ago.

When Byzantium finally fell to Islam,
Europe had already passed the Middle Ages, the Renaissance was reinvigorating men's minds, and the European expansion over the world would proceed for the next four centuries.

As a result,
North America was shaped into a big North Europe; South America provided the Iberian civilization with ample room for expansion; Africa was conquered piecemeal; disunited India bore the ruin of her fabulous handicraft industry as it was made a market to imported wares and a land of hopeless starvation. Having more solid national structures, China and Japan could not be conquered by any European power.

The fever of expansion and two world wars over, Europeans set out to prosper at home. This posed the old problem of unification, for Europe divided was too weak, especially when dealing with big federated powers.

Through their history of wars
and ceaseless change, however, Europeans revealed a great vitality. As the 1945 political division into East and West Europe turned to disappear, even more change was to occur over the hyperactive small continent.

Same as if lying on some other planet,
China was ignored in Europe until the Polos depicted her as a terrestrial country. Reciprocating, the Celestial Empire ignored the peoples living where the sun died and thought of them as barbarians.

Few Westerners ever trod Chinese soil.
The isolation of the big country was such that the national culture developed independently for nearly four millennia.

Chinese pictograms, ideograms, and phonograms reminded of Egyptian hieroglyphics for their slow evolution, but the Chinese script survived as the oldest in the world.

China's ancient culture
did not kindle the passionate interest Egypt's did in Western scholars, although the Chinese had been highly refined artists.

Planet Earth might look tiny from space,
but from the middle of the Pacific, it might not. A hemisphere of water set old China far apart from the New World.

The long Chinese history of isolation
did not end with the Manchu dynasty. The first Chinese republic was plagued by instability and underwent foreign occupation, but on becoming the People's Republic of China, the big land was isolated again—for ideological reasons.

It was improbable that Japan would have ended her long period of seclusion without the U.S. diplomatic mission of peace, amity, and commerce undertaken by the middle of the nineteenth century.

Yet, barely a century after the end of isolation, Japan had adopted much of the Western traits of productivity and aggressiveness. Determined to be the dominant power in Asia and an equal to the Westerners, Japan forced her way to industrialization by intensive study of the technical sciences and by harnessing agriculture and land taxes to build up industry and a military machine.

The outcome of World War II made all the Japanese might shift into business. The new Japanese industry imprinted a seal of bold improvement on its products, which accounted for their amazing worldwide success.

Postwar Japan wanted profits to overcome the chronic national deficit and acquire financial power. The same patriotic devotion displayed in war was revealed in business, with Japan as the sacred cause.

Japan's competitors were not only to face an able and thrifty contender, but the Japanese government of old, building trade barriers to competitive and non-vital imports. However, Japan joining the ranks of rich nations aiding the undeveloped, presented a new national image to be projected through the oncoming times of change.

When the boreal continents
turned inhospitable because of the cold of glaciations, Africa enjoyed a stimulating climate. Men emerged there and went through the finishing stages of evolution toward fully human characteristics.

The Ice Age over, Africa went tropical
as it shifted south and separated from Europe. While animals continued to prosper, the human groups not having emigrated suffered from the unmitigated heat. The excess of vertical solar radiation deeply darkened their skins, and the race lagged in a virtually endless Stone Age.

The emergence of Middle Eastern
and Western civilizations only brought Africans more misery. Lighter-skinned men resorted to Africa in search of black men, to capture them like animals and sell them as slaves. The trade and cruel exploitation of African slaves was to be forever an indelible shame on lighter-skinned men.

The many new African nations
emerging after the tragedies suffered by their ancestors looked to the future with few prospects of realization.

It was the moral duty
of all the societies having profited from African slave labor to help the Black nations acquire living conditions compatible with those of people living elsewhere. To lend them money at spiraling interests was even more evil.

The immense African crescent deserts
and retreating tropical forests were related aspects of the relentless process of desertification turning the earth back into the dry sphere it had been before water covered most of it. As forests and prairies stopped advancing by the end of the Mesozoic, deserts started to advance.

The desert belt stretched from the Sahara,
on the western coast of Africa, to the Gobi, in East Asia. The dry belt widened at a rate of three feet a year over adjacent areas along its length of seven thousand miles.

The natural desertification process
was caused by dry air, which made climate harsher, with hotter days and colder nights preventing rainfall and fragmenting rocks by abrupt temperature fluctuations. Rock fragments were then pulverized and spread by wind action to complete the making of dusty deserts.

Apart from natural causes,
the depredatory presence of men, marked by the destruction of green areas for irrational timber exploitation and boundless urbanization, helped make all lands drier.

Reversing the decaying trend
to make the green of plants advance over the deserts was a hard task awaiting many nations of Africa and those of other continents, on the common purpose of surviving.

What was remarkable
about the establishment of the United States of America was its predetermination. The people who left home to build a nation in the wilderness were different from those making a fortune elsewhere and enjoying it at home.

The pioneers were not soldiers,
adventurers, or missionaries. They were simple-minded people looking for peace and liberty in a wide open land.

Because they were distinctly motivated,
they created a different nation. The differences between the very nature of the U.S.A. and that of other nations that gave history a meaning were many and great.

The people bound for North America
knew why and where they were going. They proceeded from a highly advanced European country and were used to enjoying more liberties than people on the Continent.

Those determined men and women
were clearly conscious of their rights as Britons, but yearned for religious liberty in times when religion was a most irritating issue in Anglican England.

That historical fact was revealing
of the importance of religion in the process of formation of the U.S.A. It was also a reason for common U.S. people to be particularly sensitive in matters concerning religion.

Other lofty political ideals
besides that of religious liberty inspired the creation of a nation like the U.S.A.

But the founding fathers
were not only high-minded idealists. They were practical men who ignored none of the human frailties. The U.S. Constitution prevented widely distant extremes, from the abuse of political power to tax evasion.

It was also noteworthy
that in the first invasion of the North American continent, people came on foot and by chance from the eastern end of Siberia, via the then-existing Bering bridge.

In those times
people ignored all about ships, wheels, writing, animal domestication, agriculture. They were in the hunting stage and barely brought some crude weapons with them.

Quite differently, in the second great invasion,
people came from Renaissance Europe, where major advances had been made in all the fields of knowledge.

Besides,
oceanic navigation, book printing by movable type, steel metallurgy, and fire weapons had given Europeans an absolute technological superiority over all the other peoples on earth.

Since the foundation
of the thirteen British colonies in North America to the Declaration of Independence of the United States, only a century and a half had elapsed.

In the next ninety years,
Louisiana was purchased from France, and Alaska from Russia. Vast territories to the south and west were ceded by Mexico. Later, Hawaii joined the Union as the first overseas state.

That astonishing growth
was accountable by the predetermination with which the nation was conceived, and also by the generous availability of land on the newly opened continent. Many other factors, like massive immigration, Black slave labor, and the timely advent of mechanization contributed to the unobstructed rise of the new world power.

A nation of so distinct characteristics
was to act in different and sometimes perplexing ways. The word *American* implied a challenge to be the first in any endeavor in which the U.S. might be engaged.

For how long the young nation
could keep her splendid vitality was to be seen through the years. By the second centennial, the wear of the eventful career seemed to have left a shade of frustration in the national spirit.

While U.S. wars
were fought in the West, the nation was invincible. But when confrontations shifted to the East, the sour of defeat had to be tasted by the champion.

Pearl Harbor, the Philippines,
Korea, Vietnam exposed the determination of Asians to fight in a way U.S. strategists had never suspected.

The lessons learned in Korea and Vietnam
were that wearying non-defensive military operations undertaken in far-off fronts hardly won popular support at home, as Hannibal had felt over two thousand years before.

The combat disposition of U.S. troops,
which had been kept since World War II, had disquieting manifestations in urban centers at home. So well trained to kill were the men that many veterans could not discard the feeling that life was of little value, and an endemic wave of crime afflicted every town in the wide land.

The bigness of the country
made everything about it big. So, the racial question was big. The financial benefits of African slave labor were minimal if compared with the cost of meeting the vast social problems poised by a prolific Black minority amid a stationary White majority.

Racial discomfort
was also on the rise with the presence of a growing Hispanic minority. Large numbers of Spanish-speaking Catholics were not truly welcomed by jealous WASP nationals.

Drug and tobacco addictions, homosexuality,
sex abuse spreading herpes, and AIDS were maladies jeopardizing the well-being of the once healthy colossus.

U.S. industry
was a sorry casualty of World War II. Foreign competition, especially from defeated-then-aided Japan, deeply affected much of Uncle Sam's manufacturing business.

The Soviet peril
added financial trouble to the richest nation on earth, as it made her spend her treasure in the development of an impressive defense system.

Aggravating the economic malady,
U.S. consumers tended to buy foreign, and the formerly great exporter turned to import more than it exported.

As an overall result,
the balance deficit led to raising interest rates in order to stabilize the dollar. This plunged oil and raw materials' prices. Ballooning interest rates inaugurated an oppressive era in the world living under the dollar's might.

By the end of the American Century,
Uncle Sam's love of victory had little cause for exaltation. Great wars to win were discarded in the atomic age. As wars to lose none would start, peace by terror proceeded.

Only the space race remained
as a contest for spectacular triumphs. But a weakened economy cooled off the impetus to fly Old Glory beyond the moon, and eager scientists were put to wait for better times.

Nations of all sizes and political trends
had a dim notion of where they were going, if at all, and the U.S. could hardly tell about plans for a golden future. Territorial expansion was closed to each terrestrial power, so that growth had to be inward. Owning a vast good land and organized under an admirable system of government, U.S. creative citizens had superb possibilities for realization.

Curbing vice and waste,
reversing environmental decay, raising competitive production, assigning minorities dynamic parts in the national effort were some of the means at hand for the United States to attain great ends, thus inspiring the world through the next centuries.

The U.S. contribution
to the advancement of civilization had no precedent in history. By example, and often by direct action, the U.S. helped the world be freer, more humane, and practical, in much of the way Ben Franklin would have liked his people to do.

Spanish America
was the land of great impossibilities everybody helped make more impossible. Unification was impossible, no matter how many international assemblies spent their time in lyric talk.

Unification being a vague dream,
other dreams were not even dreamed. Many local centers of power had local interests and no vision to lead Spanish America to her role as the second power in the West.

Far from an oddity,
the dispersion of Spanish America into eighteen separate states after gaining independence from Spain reminded of Europe on the fall of Rome.

One reason for both dispersals
was that neither Rome nor Spain prepared their provinces to form continental federations of nations. Besides, Rome and then Spain were turning to decline when the peoples of the north set out to claim their share of destiny.

Among other evils,
new religious trends enfeebled the Roman imperial body the way liberalism dissolved Spanish-American loyalty to Mother Spain.

On the ground, while Europeans formed a mosaic of languages, religions, and ethnic groups, Hispanic Americans were mostly Catholic, and all spoke Spanish.

Aborigines in Spanish America
survived on the plateaus and among the mountains as farmers and mine laborers. In the jungles they kept existing as nomadic hunters. In the temperate austral countries, aboriginal Americans were extinct, while Blacks were a rarity as they were not in France or England.

Homogeneity was a poor binder
to make Spanish Americans unite. But the great disuniting factor was territorial dispersion.

Lacking a geographical center,
the long string of countries ran north-southeast along the three continental sections, from Mexico to Argentina.

The only two countries
running parallel to each other were sharply separated by a steep range of the Great Cordillera. So, Argentina and Chile were more independent from each other than they were from external influence.

The former Spanish provinces in America
had been left utterly unprepared for
self-administration. Besides, for practical reasons they had belonged to separate vice-royalties and other minor administrative centers. Prohibited from communicating and trading among each other, they had been prevented from any idea of unification.

The thorough Hispanization
of what remained as Spanish America, roughly a quarter of the whole New World, resulted from the systematic exclusion of non-Spanish immigration established by Imperial Spain.

Another cause
for the exclusive Spanish predominance was the striking innocence of the natives, who were rapidly dominated in some areas. In others, they were plainly exterminated.

Of mostly Mongolian descent
and improperly called Indians, the primitive invaders occupied the entire New World, from Alaska to Patagonia, in successive waves that proceeded as long as the Bering bridge existed.

The Bering bridge gone,
those Asiatic Americans remained isolated for nearly two hundred centuries. In the great wilderness, they advanced very little from their elementary cultural degree.

Like Africa and Australia,
America was surrounded by oceans that were uninviting to seafaring. Its regular contour, especially at the southern hemisphere, presented no great gulfs, peninsulas, or inland seas. The frightening oceans obliged to take refuge in the immensities lying inland.

Only in the last tens of centuries,
men from emerging civilizations in the Old World initiated maritime navigation on board of rudimentary craft.

Unknowingly,
carried off by winds and streams, early navigators reached the hidden continent. Via the Azores, some were left stranded on North American shores. Via Cape Verde, others ended up at the rims of desolate northern Brazil.

Unable to sail back,
they were swallowed by the vast wilderness. Vestiges of their presence were to be found occasionally, raising great controversy and conjecture.

Navigators from North Europe
later touched North America via Greenland. They called Vinland a cold land they ignored to be part of an immense continent that almost connected the North and the South Poles.

Navigating the Pacific Ocean,
adventurers from East Asia were more fortunate. Driven by eastward winds and currents, they discovered the extended Polynesian world of islands. Sailing still east, from Easter Island, they reached the eastern limits of the vast ocean on the arid coasts of Peru, in South America.

Similarly, other parties sailed east,
from the latitudes of the Hawaiian archipelago. Also reaching the eastern ends of the Pacific Ocean, they disembarked on the shores of Middle America.

The Pacific coasts of the new continent
were hospitable. Unable to sail back, or establish contact with their Polynesian bases, they all stayed. The aborigines had never seen strangers and welcomed them as masters.

Simple themselves,
the newcomers brought little progress to the native population. The civilizations they founded in several areas of Peru, Mexico, and wherever they happened to land grew ignorant of one another, as each was isolated in the immense wilderness.

According to the models
left far behind in Asia, the social organizations they established roughly corresponded to the Bronze Age level. Autocratic and theocratic, they all worshiped the sun as the top divinity.

Any notion about writing
the alien masters and their descendants could have had was jealously kept from the natives, so as to preserve their ancestral ignorance and docility. Only religion and the cult of death were inculcated.

What had really been permanent
since the lone New World was first invaded by humans was warfare for land possession. Weaker tribes were pushed eastward to the jungles, or southward to the Patagonian end of the continent.

When the Mexican and Peruvian empires arose,
more than a millennium before the European arrival, wars were already being fought with bronze weapons, although Stone Age axes and clubs were still favorite.

The abundance of gold
all over the American empires had an application in liturgical furnishings and luxurious ornaments for the ruling classes.

Aborigines themselves
did not consider gold of much value. What was noble to them was the produce of the land. They developed maize, beans, potatoes, peanuts, cacao, tomatoes, and excelled in the knowledge of the medicinal value of countless native plants. Aboriginal Americans well deserved the title of benefactors of mankind.

Over the vast expanse
of sparsely populated land, life had none of the problems of progress. Time had simply stopped. That was the world Spain found where the Atlantic Ocean ended to the west.

Thenceforth, however,
in a few centuries of European restlessness in the New World, the U.S.A. and Brazil emerged as big countries with widely different projections in history. But the scattered Spanish-American countries could hardly project anywhere.

Paying dear for disunion,
Hispanic Americans lagged behind. Their inability to overcome poverty, ignorance, political and financial instability, and administrative corruption was manifest in view of the advances made by nations elsewhere.

Typically Spanish-American,
Mexico was a member of a family from which it was separated by distance—alien, though similar. Mexico had no problems and little trading business with the republics southeast from Tapachula.

As a North-American country,
Mexico had all kinds of unfavorable relations with her northern neighbor. The close presence of the U.S.A. was overimposing to Mexico, and so it was to Canada.

The long unfortified Mexico-U.S. border
in later times was not violated by punitive expeditions either way, but by myriads of Mexican illegal immigrants.

In the opposite direction,
U.S. tourists invaded Mexico for a taste of cheap good life and freedom from regulations, but few remained. Mexico would not turn Anglo-Saxon, but the U.S. population was becoming Hispanized to an unpredictable degree.

Less inhibited
by social and financial considerations than the Anglo-American society, Mexicans procreated freely. The great population surplus had a long open border to reach a promised land. After all, the U.S. bordering states from the Gulf of Mexico to California had for long had a native Mexican population.

With a long common border with the U.S.,
ample coasts on both oceans, and only tiny Guatemala as a southern neighbor, Mexico was a fascinating country. Notwithstanding her Indian ancestry, Mexico had more of a New Spain as for character and culture as a result of three centuries of intense Hispanization.

As it could have hardly been otherwise
with an astronomical unfavorable balance, Mexico was deep under U.S. economic influence.

Yet, differently from Franco-Canadians,
and for all the U.S. immediacy and influence, Mexicans living either at home or in the U.S.A. were barely receptive to Anglo-Saxon ways and language.

And it was only true
that the Mexican minority pervading populous areas of the Union, together with other Hispanic minorities, was straight in its course to divide the U.S. identity into one Anglo-Saxon and one Hispanic, Mexicans being reluctant to participate in the national life of the guest country.

Still, few Mexicans
planned to prosper at home. Rather, very many dreamed of going north to enjoy a prosperity they were unwilling to build in Mexico. Persistent inflation was no help, anyway.

Central America was a victim
of continental drift. In a hundred thousand years, it was to disappear, crushed by South America drifting north to form a single land mass with North America.

The string of small republics
had problems derived from its geographical configuration. Five separate countries consistently failing to confederate succeeded instead in preserving their independence from each other, if not from strong outer influence.

As a former department of Colombia,
in South America, Panama had a separate history. Panamanian problems were not related to Central America, but caused by the isthmian position of the country.

Being more feasible
to cut an interoceanic canal in Panama than elsewhere in Central America brought Isthmians plenty of trouble, starting with their unfortunate secession from Colombia.

As a Colombian department,
Panama had been economically more important than it was later as a small separate republic.

Cut in two, physically by the ditch,
and politically by the U.S. Canal Zone, paid a modest rent as the Isthmus landowner, tiny torrid Panama sustained an epic struggle for her rights. But as any entity building the canal, either a company or a country, had to exert too much power over Panama, friction and trouble had to be permanent while the canal was being operated by that entity.

For the rest, hot-blooded Panamanians,
as all Spanish Americans, had plenty to lose in their normally unfavorable relations with advanced countries.

Mountainous Costa Rica,
enjoying a pleasant climate and having nothing of the African-Jamaican population introduced to Panama for the canal works, was a pretty land of limitations. Confined by two oceans and two sultry neighbors, the land suggested the romance of Shangri-La.

Nicaragua was the second choice
for the opening of an interoceanic canal, utilizing the Lake of Nicaragua and the San Juan River. One of the warmest countries in tropical America, Nicaragua underwent a long era of dictators. Their suffering not over, Nicaraguans still had a hard time ahead only to gain the right to live in peace.

The only Central American country
with no Atlantic shores, small overpopulated El Salvador told a story of acute social injustice. Eventually, the republic became one of the most restless spots in Spanish America.

Honduras, homeland of Francisco Morazán,
the gallant hero of a lost battle for Central American unification, did not suffer the grave tensions afflicting her two southern neighbors. In the still thinly peopled country, there was room for Hondurans to prosper—unless the general Middle American situation turned much worse.

Guatemala was a showcase
of natural beauty. Lakes Atitlán, Amatitlán, Izabal were gems set in the green of mountains, inviting to contemplation. All suggested that natural agents sometimes joined forces to create sceneries that could be described as superb, even by visitors from perfect worlds.

Unfortunately,
the very agents creating that physical harmony were ruthlessly destructive, and Guatemala was a special victim of earthquakes.

The still emerging areas of Antillia,
a half-submerged continental mass projecting from northern South America to the peninsula of Florida, were forty densely populated islands known as the Greater and the Lesser Antilles.

The largest of them all,
Cuba had been a Spanish province up to the U.S.-Spain war of 1898. Defeated Spain then ceded Cuba, Puerto Rico, and the Philippines. Thereupon, Cuba, even as a republic, continued to live under strong foreign influence.

The Dominican Republic had a hard time,
especially due to the French occupation of the Haitian part of her island. Having returned to belong to Spain twice, and her proposal of annexation having been rejected by the U.S. Senate, the small republic was left to build her own future.

Haiti's African population, introduced by France,
had a long suffering under a succession of dictators who kept the beautiful hot country as an example of poverty and despair.

Puerto Rico was retained by the U.S.
as an associate state for her strategic worth as the "Gibraltar of the Caribbean." Of the Lesser Antilles, Trinidad-Tobago was a substantial oil producer and enjoyed a comfortable situation as a member of the British Commonwealth.

South America was divided
into two areas of diverging perspectives. One of them was Brazil, of Portuguese ancestry, lying over nearly half the continent and solidly unified. The other was nine Hispanic republics planning anything but unification.

The northernmost of the nine,
Venezuela and Colombia, at first forming the Gran Colombia with Ecuador included, were prompt to separate.

Venezuela was a major oil producer
with incredible financial problems in spite of an immense wealth that was the gift of nature, and not the product of determined work.

Home to chronic violence,
Colombia spent a substantial part of the national treasure and energy in fighting crime and subversion.

The finding of petroleum
in Ecuador somehow expanded the limited national horizons, but petroleum, as all exhaustible resources, was a fickle promise of prosperity.

Suffering from political instability
and inflation, the endemic Spanish-American maladies, small Ecuador had to face the times ahead without much help from her Gran Colombian sisters.

Peru bore the weight
of archaelogical riches that were irrelevant to Peruvians
of republican times.

By the time of the Spanish conquest,
several nuclei of isolated civilizations founded by
stranded navigators had already perished.

The last of them, however,
had grown and was extant. It was the Inca empire,
which was easily submitted and destroyed by the
Europeans.

Peru presented three sharply different areas,
none inviting the advances of substantial development:
the sandy deserts of the whole coastal area, the frigid
Andean sierras inhabited by destitute aborigines, and
the rain forests forming part of the Amazon basin.

Suffocated by an exploding population,
crushed by external debt, illiteracy, social injustice,
political confusion, irrepressible subversion, Peru
seemed to lack the vitality of young nations.

Rich in natural resources
and favored by a splendid geographical position, Peru
nevertheless had enviable material possibilities to be a
prosperous and joyful country. All it needed was spirit
and leadership.

Argentina was the one Hispanic country
in South America with significant prospects, plus the finest metropolis in the Spanish world. Strangely enough, the nation revealed disquieting signs of weariness and declination.

Sharing with Australia
the melancholy affecting folks living in austral latitudes, Argentina was a land of pessimists who scored negative rates, both in population growth and economic expansion despite the food abundance yielded by an extraordinarily fertile soil.

Quite different from Mexicans,
who treasured their Indian past, Argentines were vain about being of pure European descent. Particularly sentimental, they were poorly motivated save for a fuss about soccer and an old issue on the dogged British occupation of some desolate isles. A sorry war of reconquest brought the nation only more chagrin.

Added to the spending and nationalizations made
by a populist regime thirty years before, the dire cost of the war against U.S. aided Britain crippled the economy of the once richest Hispanic country, causing an inflationary process that, if not firmly checked, was to impoverish ensuing generations of Argentines.

The Argentine melancholy and aloofness,
together with the Chilean isolation, made the tapering end of the continent very lonely, indeed.

Uruguay's history of struggle
was disproportionate to her size and population. The smallest Spanish-speaking republic in South America had to pay a high price for independence, only because of her position on the left bank of the Plata River.

In the years following
the Spanish domination, Portugal, England, then Argentina and Brazil at one time or another had a stake in the strategic Cisplatine Province.

Her first class geographical position
would have placed Uruguay either as the second important Argentine province or as a Brazilian state rivaling São Paulo. As a buffer state, Uruguay was bound to a rather modest destiny.

Paraguay would not have lived
under durable tyrannies had it been a mountainous land. From the Greeks and the Swiss to the Afghans, mountaineers had a name for liberty lovers, as mountains nurtured individualism.

Paraguayans were a frugal stock,
one of whose dictators engaged them in a six-year war against three nations, two of them being the largest in South America. Mediterranean, hidden across their neighbors' wildest margins, Paraguayans had a real task ahead to recover time and freedom.

Bolivia epitomized
the adversities befalling the Spanish world. Having fought for fifteen years to gain independence, the republic lost two-thirds of her original area in wars with encroaching neighbors.

Still, the country kept far more land
and natural resources than could be readily utilized. In other words, Bolivia had reserves for the future.

Shortly after the Spanish conquest,
Bolivian silver mines supplied the wealth Spain needed to assume a leading role in Europe. Bolivian forests yielded the rubber for the tires on which the newly born auto industry made its way to prosperity.

Bolivian par excellence
were potatoes, which acquired German and world citizenship. And because of the astounding inability of nations to suppress cocaine addiction, while synthetic substitutes were being developed, Bolivian coca leaves and those raised elsewhere caused a faceless, ubiquitous financial power to grow, fed by prices ballooning as a result of the ban on drugs.

The Bolivian landscape
amounted to one of the most impressive on earth. A chain of some of the highest Andean peaks melted its snow to fill the Amazon and Plata basins.

Bolivia was structurally mediterranean. Although near the Pacific Ocean, the country was isolated from it by mountain walls rising from abysses and trenches opening in the ocean bottom. The Bolivian coasts on the Pacific were awfully dry deserts that happened to be full with mineral riches. Amid the deadliest desolation in the world, those riches were hard to defend against disembarking marines, as were the developments at the 1879 Chilean task force operations, also performed over Peru.

Cut from the Pacific and far from the Atlantic, without navigable access to either the Amazon or the Plata, Bolivia was quite inland. The aboriginal majority speaking native languages would hardly be incorporated in the national life. Certainly, American Indians had been living on the continent for millennia before Creoles even existed.

The Bolivian plight of underpopulation, illiteracy, and the ignorance of many able to read was aggravated by the lack of patriotism in too many of the government. By the national institution of graft, officials every day grew richer, while the people turned only poorer.

The prospects of most Hispanic American republics of surviving as independent states through the times depended not so much upon themselves, given their disunion and inertia, but upon currents of thought permanently altering the world balance of power.

The solid geography
of the vast Portuguese possessions in South America helped their unification into the Brazilian Empire, which soon became the largest one-piece federated republic in the New World.

Whereas Yankee pioneers and settlers
pushed westward by themselves, Brazilian bandeirantes—and the Portuguese army—advanced from the Atlantic coasts, wresting as much land as they could from the wilderness, from primitive aborigines, and often from Spanish possessions.

So, before the Portuguese rule had ended,
Brazil was already the largest, most motivated, and aggressive country in South America, with still unquenched ambitions, like that of reaching the Pacific Ocean.

The immense Amazonian hinterland, though,
offered more frustrations than possibilities. Tropical South America was not so friendly to human advancement as Europe had been. The primary soil of the Amazonian forests was not fit for agricultural development. The northeastern wastelands only yielded legions of desperate migrants.

Largely submerged in the tropics,
Brazil was burdened with high rates of population growth and unemployment; inflation being a national calamity. Without attaining financial order, the Brazilian capability of standing as a first-rank industrial power was to be gravely hampered.

The lands of men
were infinitely varied, and each nation seemed to be taking a different course.

At the same time, however,
there were powerful forces tending to equalize the problems, needs, and aspirations of human groups everywhere.

Such community of interests,
added to rapidly improving communications and heightened levels of general education, was taking mankind to a habit of reaching accord, even on formerly absolute differences.

X. – North-South, East-West

The North was up and the South was down.
There were complex physical differences between the top and the bottom of the world. Some were of electromagnetic nature. The earth's iron core acted as a huge magnet, inducing the electrical phenomena generated by the atmosphere.

The polarity of the cosmic radiation
attracted by the North was thus opposite to that drawn by the South. Subtle contrasts in certain biological processes were to be found among others in different orders.

Toward the North Pole
continents crowded together, surrounding a frozen sea. The northern magnetic pull seemed to act on the drifting continental plates in addition to the inertial effects of the earth's spin on the fluid mass of the oceans.

The South Pole
was a super-cold continent topped by an abrupt mountain range. Much of the lowlands lay under sea level, all being covered with the thickest ice sheet on the planet.

Antarctica was being deserted
by the southern continents and their pointed ends loomed more distantly as time passed. The southern magnetic force seemed to be repelling, so that eighty percent of the world's lands lay north of the equator.

The way continents drifted north,
there was a new Pangaea to be formed surrounding the North Pole, with only Antarctica left, frozen at the bottom.

Broadening upward,
the lands north from the equator were more densely populated. Contrarily, the southern continents tapered off downward. The languid landscapes of austral South America were being deserted by the few still staying. Aborigines of Tasmania, the lone island south from Australia, simply died out.

Any point below another was south,
even at the highest latitudes. The only northern site with no south in itself would have had to be a sharp central point in the North Pole, but the spot could not be located with the earth's axis tilting constantly.

One reason for northern summers
to be hotter was that there was more land to the north. Accordingly, Antarctica, surrounded by water, was much colder than the North Pole. Related factors determined the beautiful coloration of the boreal foliage in autumn, while to the south withering leaves turned only gray.

As a geometric line of reference,
the equator cut meridians in two. More than sharply separating both hemispheres into north and south, the equator and both tropic lines marked the warm belt of the planet.

As a climatic factor
influencing human behavior, the northern chill imparted discipline and industry in human groups thriving far from sunny areas, provided white winters alternated with green springs. That was a special reason for the progress of some boreal nations.

Yet, at far higher latitudes,
there was no alternation of the beauty of seasons; only half year long days and half year long nights. It was so cold in the polar regions that people were few and primitive. Arctic Circle Eskimos were about as innocent as austral Fuegians nearing extinction at the cold end of South America.

East and West
were less objective and more relative geographical notions. East and West only suggested the direction of the motion of rotating planets with respect to the sun. Horizons where the sun rose were east, and west were those where the sun set.

A non-spinning planet
would have no east and no west, although it would still have a north pole and a south pole in relation to its orbital plane. And when the body rotated clockwise, as Venus did, east was west and west was east.

The idea that China was east
could be European or African. From America, China was west. To a Chinese, Europeans were western, but people from the New World were really eastern.

The sun emerged from the East every day.
Civilization also emerged from the East, once in history. By the time Orientals were refined artists and statesmen, Occidentals were learning elementary farming.

From Asia to Europe,
then from Europe to the New World, the route of human advancement was westward. And west it kept moving as U.S. technology crossed the Pacific to revolutionize Asia.

The Western life-style,
with its complex materialism, environmental pollution, stress, deadly weapons, vice, was eagerly assimilated by the Asians, and it deeply altered their old ways of life.

There never was a serious Yellow Peril
to the West, but the White Peril was quite tangible to the East.

XI. – Dominance and Dependence by Technology

Primarily,
technology was the science of technical arts devoted to saving physical effort and amplifying the power of muscles by the use of available forms of energy.

Animal traction applied to hauling loads,
winds pushing sailboats, river streams driving grinding stones expressed in facts the distance starting to separate civilization from natural life.

Up to the first decades
of the nineteenth century, however, there had not been much advance in the mechanization and the complexity of life.

Except for ore smelting and alloys,
raw materials underwent little transformation. Food was largely of local production. Imports from overseas were limited to a few items, given the time taken and the risks run by sailboat oceanic navigation.

As the nineteenth century advanced,
progress accelerated and the bases of a technological era were fully set up. Motors and engines passed from being the objects of endless experimentation by eccentric inventors to become indispensable to civilized life.

Communications were accelerated
from the pace of couriers to the speed of cablegrams. Steamers drove industry to develop into an economic force, and air pollution began to distinguish industrialized nations.

Then, electricity
and internal-combustion engines gained the world. Food and goods production soared. The world population exploded, thereby getting ready for the greatest carnage ever.

World Wars I and II
would not have been possible without engine-driven tanks, submarines, aircraft. The blitzkrieg was essentially motorized.

Motorization
aided logistics powerfully, so that wars could be fought at a larger scale than ever before.

Peace was equally motorized.
Without the engine, untold millions would not have been killed in auto and plane crashes.

Sulfurous contaminants
given off by coal-burning steamers readily combined with carbon oxides from engines burning petroleum incompletely. Nuclear fission of uranium added radioactivity to an environment plainly turning lethal.

In a number of urban generations,
transport facilities were to impair people's ability to walk. Human legs, which had taken their proprietors so far along the routes of migrations, were to be employed less and less.

Poor blood circulation
from lack of exercise, as the cause of heart conditions, was responsible for a universal bloodless massacre.

Like a mighty blind genie,
technological progress dispensed blessings and curses alike. And the Aladdins having called the genie seemed unable to make him undo the discomfort and misery brought along by the wide use of machines.

Either a wonder or an evil,
mechanization was not a creation of every human race, but once created, machines promptly became universal.

Left alone,
the large majority of men would have kept their habits while living conditions were not inviting—or compelling—to change.

Still, a disquieting impulse to change
the old ways and go beyond the familiar valley could arise any time from imaginative individuals who kindled their passion in others.

Renovating trends were contagious passions against which the bulk of mankind, although deeply conservative, lacked effective defenses. All the great improvements in the quality of life came from the courage to change, either spontaneously or by remote inspiration.

Change might occur in widely different orders of life. Great behavioral changes were inspired by philosophers seeking to elevate the moral standards of their fellow men. Many renovating drives took place in East Asia and were superseded by new ones when worn out by habit or relaxation.

Great changes forward, brought about by pure thought and scientific investigation, were characteristic of the peoples living west of the Urals.

In a world without the inventiveness and enterprise of the young European civilization, it would have been hard to imagine scientific-minded cultures emerging among Orientals, who had rounded their historical cycle, or from primitive Africans, or even from isolated American aborigines.

In its impelling reality and convenience, the Western tempo of material progress became irresistible, and the world only followed it. Railroads were built in India as they were in China, Africa, and America.

The Westernization of the world
entered as an epochal chapter in the history of civilization. The life-styles of civilized people, like that of traveling, were modified as the mechanized times advanced.

Technical sciences took the old sport of war
to the nuclear stage, where contenders could exterminate each other. The rules of war thence had to be rewritten.

To go or not to go industrial
was no longer a matter of national character. To a simple nation, not to have an industrial structure spelled high prices for manufactures and low revenue for raw materials.

Still, with determination and purpose,
undeveloped countries could possibly upgrade themselves by encouraging inventiveness and initiative among their own citizens, the best stimulus being financial reward.

The Western lesson of industriousness
was clear, and industrial powers began to emerge outside Europe and the U.S.A., establishing that high technology was not the exclusive domain of its initiators.

The changes caused by machines
in the structure of civilization had merely started, but had already altered the trends of culture everywhere. The art movement reflected the anguish of an era of transformation and rupture with the past.

Music was everywhere and always.
As the oldest of the fine arts, music was a world in itself, and the realm of machines was at the service of beautiful sounds. The technics of recording and reproduction put the classics to the enjoyment of every refined soul. Popular music filled all the moments of daily life.

Since the invention of music notations,
composers gained the acknowledgement they had for long deserved. Among them, many geniuses became the favorites of the noble and the rich. Later on, as the music recording industry prospered, popular musicians commanded fantastic royalties for the millions of recordings sold.

The evolution of architecture
ran parallel to that of clothing. When men wore no clothes, they built no houses. Then, plain clothes corresponded to modest housing; ornamented robes to palaces.

Successive civilizations could be identified
by their architectural styles. For millennia, the pyramids made visible the Egyptian passion for duration. Their heroic sizes gave proof of the greatness of the Old Empire.

Babylon's hanging gardens
soon were mud again and taught that architecture without stone lasted nearly as words spoken and not written.

Around the Mediterranean
the Romans built roads and aqueducts that were to be good even for the next era. Truimphal arches throughout the Roman world depicted the deeds of emperors and their legions.

The needles of Gothic cathedrals
pointed to heaven, in a refined expression of medieval spirituality.

Renaissance architecture and related arts
took another course, returning to the solid structures and pagan tribute to nudity of classic times.

After the Gothic style
and the neoclassicism that followed, no major trends in architectonic aesthetics beautified the West.

As the richest country
since the nineteenth century, the United States jumped from the neoclassic lines of many government buildings to the tasteless reticular geometry of commercial skyscrapers.

The spectacular skyscraper style
was initiated when steel frames were applied to support very high towers for practical use. As in other U.S.-made things, there was no art in skyscraper architecture. Its cold beauty came from the economic might it suggested.

How the noble arts would survive
the ravages of so much transformation depended on their power to fill a place in the minds and the spirits of people rushing after tangible satisfactions.

Painting seemed to be a casualty
in the clash of the fine arts with the practical sense of material progress. After prolonged centuries of anonymity, painters enjoyed a period of glory opening with the Italian Renaissance and closing when photography reached unobjectionable perfection.

During the bright four-century cycle,
the switch from frescoes to oil paintings was a vital advance in plasticity and durability. The Western pictorial art was enriched with a profusion of wonders executed by geniuses who would have no successors.

The perfection of color photography
suppressed the need of a brush to put in pictures the sights of life. The eccentricity of cubism, impressionism, and surrealism, though daring, only prompted the decadence of painting.

Painting and drawing
thenceforth passed to be auxiliary crafts in the technics of illustration. The field of depicting imaginary subjects, and those real but impossible to photograph, was still open to persons with imagination and the gift of handling a brush.

With painting relegated to auxiliary functions,
and sculpture back as a statue-making trade in charge of adorning parks and gardens, the plastic arts remained still, depending on photography for appreciation by distant viewers. Masterpieces, besides, were prey to vandalism, theft, decay.

The stillness of the plastic arts
contrasted with the dynamics of literature. Writing styles evolved freely to deal with what occurred or would occur, really or fictionally, for written words kindled the imagination more intensely than speech or images.

The impact of meaningful writing
went beyond the limits of language or time by translations from any decipherable script into any other language. Read aloud, the message of inspiring thoughts and original ideas would impress even listeners unable to see or read.

Literature was the youngest of the fine arts.
It had to wait for the invention of writing, as oral accounts by heart were fickle at best. Then, manuscript books were treasured as gold, even chained to desks at medieval schools.

The early advances of typography
promoted literacy greatly. Primitive printshops using movable type put out in years more books than those hand-copied in a millennium. Initiating the movement toward a universal culture, the Enlightenment was a product of literature plus technology applied to the typographical arts.

Every improvement in the printing methods contributed to the quality of reading material. Issued by the millions, splendid books and magazines were available at low cost for buying, and at no cost at public libraries.

The diffusion of education and culture by the mass production of books was to change the world intellectual trends in a measure that not many could foresee, even in view of the facts that books were of little use to the illiterate and that unrestricted reading was not permitted in a few countries still under dogmatic regimes.

That greatly enhanced tendency to change forward was of paramount significance in order to consider that technology aiding to spread knowledge was a driving force of human progress.

XII. – The Demise of the Weak

Wherever life prospered
there was competition for food and living space. Competing species coexisted until one prevailed over the other. Flight or submission, when not death, was the lot of the losers.

Competitors might belong
to different zoological orders, but would often descend from common ancestors.

The weak stayed among the living
only for short. Besides competitors, the ever-changing medium and deficient organic defenses made survival problematic for many that would have liked life easier.

Once the only homo genus
surviving from a family that had surpassed animaldom stood out as the only human species on earth, no other kind of creature was left around to defy human predominance. Great wild beasts died out or hid in the wilderness.

Struggle persisted all the same
between human groups externally differentiated by climate or simply warring over land. Since the emergence of man proper, however, no human group as a whole seemed to have been exterminated by another, despite large-scale massacres.

Aborigines were partially exterminated
over the plains of North and South America. Unlike docile mountain Indians surviving in large numbers, those from the open plains were too fierce to be put to forced labor.

Ironically, tobacco and coca leaves,
originators of great hazards to the health of people, were first raised for different purposes by American Indians.

In North America the natives smoked tobacco
through long pipes as part of ceremonies. Andean natives chewed dried coca leaves to kill hunger and pain. But latter users became compulsive cigarette smokers and cocaine addicts, thereby inflicting themselves mortal physical damage.

The Western Indians' tobacco-coca curse
together with the opium-hashish Eastern curse struck many of the weak everywhere, though some victims would not have liked to think of themselves as weak.

Since antibiotics saved lives by the billions
and an indefinite peace by atomic terror stayed on, deadly vices spread as if to check the numbers of the weak and the unmotivated. Sexual diseases and homosexuality helped greatly.

Self-destruction by stimulants
was committed at great cost to the victims and at big profit to the suppliers, as the ban on drugs caused drug prices to rise astronomically.

The financial success
of the drug business, for all the harsh and sometimes bloody enforcement, was a perplexing development of the years following the dollar recovery upon the petroleum crisis of the late seventies.

The main cause of that development
was the fast-growing demand for stimulants, which gave an appalling notion of the frailty of victims to reject poisons.

Another cause
lay in the artificially high prices insuring an uncontrollable drug supply and the accumulation of fabulous tax-exempt fortunes by big drug dealers.

Meanwhile, the misery of the weak
remained unmitigated. No scientific-minded plan was ever launched to make stimulants unappealing, or at least less lethal.

Certainly, through general agreement,
all nations were capable of ending the unwinnable dirty war on drug suppliers simply by submitting stimulants to more severe regulations and taxation than those applied to other hazardous substances like tobacco and alcohol. Greatly reduced drug prices then would have tacitly eliminated crime and corruption without necessarily pushing up drug consumption.

Governments everywhere
persisted in the conventional approach to the drug problem because of a general lack of insight into the anguish of the times and the complex psycho-social agents thrusting the weak into vice.

There was also a wide indifference
about the economic factors created by the abnormally high drug prices, which fomented greed in too many, precisely when the prices of many other commodities were plunging and unemployment in undeveloped countries was rising hopelessly.

Everything led to the conclusion
that the worst social tragedy ever was waiting too long for the world officialdoms to act sensibly.

XIII. – The Ways of Universal Life

When the knowledge of the physical conditions leading to the emergence of life were still vague, popular imagination conceived fantastic manlike creatures populating the moon, Mars, Venus, and even the Sun.

Time passed,
and when it became apparent that the neighboring planets were lifeless because they lacked water, science fiction resorted to imaginary worlds elsewhere.

Astronomical science popularizers
lectured on exobiology based on elements other than water-based cells. Some even radioed signals into space, awaiting response from civilizations in the cosmic vicinity, obviously unaware of the high degree of dispersion of universal life.

The question of life
prospering beyond the earth at all levels of simplicity and complexity, however, was not out of the domain of disciplined reason.

On the premise that life on earth
developed under adequately combined and durable physical conditions, it was sound to reason that such special conditions seldom met, large as the number of planets circling the stars throughout space could ever be.

To establish the existence
of life manifestations in distant worlds was to be far from the possibilities of any civilization, whatever its degree of advancement, for the absolute reasons of distance and time.

The distance separating stellar systems
could be understood only in figures of light years. Such abysses were impossible to cover during the lifetime of living creatures, considering the slowness of inertial bodies traveling through space as compared with the speed of radiant energy.

The time barrier was also absolute.
Given the immense differences of time between the formation processes of widely separated life-abiding solar systems, by the time a civilization emerged in a remote cosmic area, others might have vanished far away and long ago, together with the worlds where they had existed.

The terrestrial mechanized civilization
was very young, and so was the human race by genetic standards. Multiplying the sum of the human progress reached in five millennia only by ten would have suggested how far advanced longer lived civilizations elsewhere could have been proportionally.

An imaginary look
at the accelerated progress of human science in the next twenty centuries was disturbing, but could give certain idea of the power civilizations in distant worlds might have built up over even longer periods of progress.

Any unlikely determination
of the existence of a far-off inhabited world would have naturally been sensational, but it could have hardly affected human life, remote as other solar systems were in time and space.

Stars needed all the space they took
in their traveling, subject to their own inertia and to complex forces of gravitation. So, the whole vastness of space was necessary for the universe to exist.

Amid the cosmic immensity, a practical objective
of a highly advanced earthly civilization had to be putting nearby solid planets under control, so as to make them extensions of the human dominion in the solar system.

As for the void of human knowledge
about life developments in outer worlds, a way to fill it had to be a more detailed conception of the marvelous story of life on earth, ever since it started from the structure of water-based vegetal cells. That very kind of universal cell structure had to evolve through largely similar patterns wherever water flowed under durable suns.

Three billion years
of uninterrupted biological evolution on earth produced an infinite variety of living forms; none of which was supposed to be final or perfect, but all able to evolve in order to follow the changes in the physical medium.

For the generous productivity
of the vegetable kingdom in its vastly varied species; for the amazing features of jellyfish, lobsters, octopuses, dinosaurs, and for the sublime beauty of women, Mother Earth should have won every universal contest of fertility and creativity.

Mindful of the earth's record,
an aware observer with imaginary powers to visit remote inhabited worlds would not have been surprised at the strangest alien creatures—small or big, simple or complex.

XIV. – The Great Challenges

To be born implied to accept
the challenge to survive. As enclosed fortresses full with protein reserves, organisms were under the permanent siege of infectious predators and parasites. A healthy individual was the image of efficient organic defenses.

Not everybody was fit to survive.
Very many died soon after birth for inborn lack of defenses. Others would lose their defensive powers at any stage of life.

The battle against the maladies
caused by invading biological agents was the central function of medicine. To enhance natural defenses was as vital as checking the microscopic enemies of human life.

No medicine cured by itself,
but by helping organic defenses in the effort to restore health. When malignant germs seeming to be normal cells or coming from new brands attacked, or when vital organs were severely damaged, the whole organism turned helpless.

Not all persons aged at the same rhythm.
Whereas some individuals aged faster, others enjoyed longer and healthier lives only by maintaining physical fitness. It was fine to live longer provided life kept its wonder and excitement.

Successful physical survival
was in itself a challenge to accomplishment, which not merely consisted in acquiring wealth and reputation.

Accomplishment was the personal record
of positive actions to help improve the quality of human life. In the intellectual sphere, accomplishment meant to widen men's horizons through education and discovery, not to submerge minds in superstition or fanaticism.

With no possible alternative,
the way to meet the challenge to accomplishment was the acquisition of knowledge. The great sages of mankind created master works to ennoble and inspire everybody. High learning ceased to be exclusive long ago.

Specialists mastering
their particular branches of knowledge were no doubt valuable and important, but by ignoring all other disciplines of the intellect, they would expose unprotected flanks, as ignorance and nakedness could be total or partial.

Opened by the progress
of communications and education, the wide access to learning made apparent to every mind that increasingly more societies aimed to advancement were to be represented by the top category of accomplishers, also known as geniuses.

In an almost wholly aquatic world,
water scarcity was an irony. It was also a great challenge, since only a small proportion of the world's water was fresh.

On a planet totally under sea water,
fresh water would have occurred but briefly, during rainfalls. Free from salt, water was only running, gaseous, or frozen. Major lakes were saltier than rivers.

There was so much affinity
between water and salt that, given the abundance of both, water tended to be salty wherever it accumulated for long. Practically, saltiness was the permanent state of water on earth, freshness being transitory.

Under that natural circumstance,
a large increase of fresh-water consumption by a thirsty civilization had to make fresh-water scarcity inevitable. Worse still, the alteration of the atmospheric composition by the increase of carbon and nitrogen oxydes from the burning of coal, fossil oil, and gas raised atmospheric temperatures and prevented normal rainfall.

Recurrent droughts scorched farm lands
and reduced food production. Elsewhere, heavy rains flooded lowlands, also destroying agriculture. All showed that, while quite able to destroy the medium, men were incapable of exerting any action toward a system of weather control.

A probable explanation
for the general passivity on the water problem could be that it was not too pressing after all. Local solutions were usually enough, like tapping underground deposits and damming rivers.

To pray and wait for rains
was the easiest solution. Dancing and chanting also helped. Every fact suggested that at a fairly advanced stage of the applied science era, men still acted with much of the early agricultural approach regarding the critical water problem.

Apart from rains, underground deposits,
snow- and ice-melting, there was no other fresh-water source. But it was visible to all that most of the rainfall on earth was the product of large-scale sea water desalination by evaporation under solar heat.

Hence, the practical procedure
to answer the challenge of fresh water scarcity was to follow the natural process. Desalting sea water along the extensive coastlines of the world was possible by using solar heat for evaporation purposes, in volumes subject to endless increase as the operating methods improved.

A worldwide pipeline network
for water distribution wherever necessary would have thus produced revenue to sunny coastal states exporting fresh water.

The challenge of satisfying
the energy requirements of a growing world population was even harder to meet. Energy crises began to affect civilization as soon as transport underwent motorization.

The vast coal reserves
lying everywhere would have been enough for a very long time with only steamers turning heat into motion. But the advent of internal-combustion engines burning fluid petroleum instead of solid coal took carbon mines into abandonment.

The sudden interest in fluid fuels
put petroleum in great demand. Somewhat earlier, electricity was being seen as a great promise. Electric motors were simpler, cleaner, and more efficient than gasoline or diesel engines.

But electricity was an artificial form of energy.
It had to be generated in plants and could not be stored or transported as fuels could. Using heavy batteries, the first electric automobiles, which were also the last, performed poorly. So, electricity was confined to stationary uses and a few systems of electric trains, especially subways.

When hydraulic generation of electricity
could not meet the demand, coal, oil, and gas were burned to drive large dynamos, oil being easier to transport, handle, and burn.

The practical advantages of using
fossil fluid fuels were decisive. As it came to be that only by burning oil civilization would go on and wars could be fought, a steady oil supply became vital to all nations.

In the search of new energy sources,
atomic energy, once hailed as the final wonder, happened to be a hazard rather than a solution. Nuclear power was hastily put to win a war, and very soon to boil water in steamers driving conventional electric generators.

Physicists' inability
to convert fission-generated heat directly into electricity was exposed, and it was again when the radioactive wastes of fission could not be safely put away or eliminated.

The hazards of nuclear fission
were inherent to the processing of radioactive fuels. Nevertheless, a system of final nuclear-waste disposal would have been to pack the deadly matter in capsules and launch them into outer space or into the sun. Several objects had already followed either course in the shape of spacecraft.

But nobody seemed to have considered
such an alternative, and the accumulation of radioactive residue, buried anywhere or dumped into the sea, was fouling the medium progressively, aided by spills of nuclear fluids from clumsily operated reactors.

The world's expectations from the atom
then switched from fission to fusion. Golden hopes were placed in a process that would yield endless amounts of energy from the hydrogen of water, without leaving lethal wastes.

As it happened,
intensive international cooperative experimentation failed to keep high temperatures constant for hydrogen plasma to fuse in a sustained reaction, simply because containers, the atmosphere, and everything around acted as coolants. Stellar hydrogen fusion proceeded for billions of years, contained only by the void of space.

Another misconception in the grand intent
was to have tried to obtain thermal energy, instead of radiation, in confined operations of atomic fusion.

Certainly,
stars were huge hydrogen concentrations reaching the fusion point by the high temperatures at the photosphere. Such temperatures were caused by the strong stellar gravitation exciting the gaseous mass. In the process, hydrogen atoms fused into helium atoms, giving off large amounts of energy.

Part of that energy
turned inside the star mass, heating it up progressively; the rest was discharged into space as radiation, not as heat, without warming up the void and without dissipating.

Scientists dreaming
of earthbound hydrogen fusion marvels distantly echoed medieval alchemists in the quest of the elixir of youth and the making of gold.

Civilization, in its restless growth,
however, called for more practical sense in the search for reliable fountains of energy.

It was beyond discussion
that whatever the means of generation, hydraulic, atomic, or any other, electricity served stationary purposes mainly, and that transport, in an overwhelming proportion, was energized by oceans of liquid fuels.

The inescapable conclusion
was that, given the transit to exhaustion and the fouling traits of fossil fuels, a worldwide scale production of liquid fresh fuels was the only means left for civilization to keep on the move.

Completely unpolluting,
fresh liquid fuels were to be produced by the sun, which was the very source of energy having bathed the earth in light since the birth of the solar system. And the agency to capture and process that energy into the calories of fresh fuels was the natural and highly efficient process of photosynthesis.

A large variety of plants produced sugars. Many others produced oils, and both sugars and oils were good for the distillation of highly combustible spirits out of their hydrogen and carbon contents.

From the selection and blend of the most adequate types of ethyl, methyl, and oil alcohols, fuels for the diverse engine combustion systems were obtainable to meet whatever increasing volume of demand.

According to climate, soil type, and particular flora, every country was to be able to produce its own fuels, if needed, under the guidance of international corps of specialists, but free from the dependence on foreign oil suppliers. Also liberated from the economic asphyxia suffered by nations lacking oil reserves.

World agriculture was to be assured of increasing prosperity by the demand of produce ranging from sugar canes and beets to oleaginous seeds and fast growing bamboos, aquatic herbs, sea weeds.

Either from organic or geochemical origin, fossil oil deposits were not inexhaustible and should not have submitted civilization to deepening environmental fouling plus economic dependence. The challenge of an ever-looming energy crisis was to be fully answered by solar power, photosynthesis, agriculture, and technologists.

Providing the world with first-rate clean-burning fuels was an added assignment of agriculture to be inevitable on the exhaustion of petroleum. Highly improved farming methods and extended cultivable areas were then to be requisite.

Extensive forestation of sloping zones and replanting of ravaged woodlands were crucial against soil erosion, damaging winds, and to restore normal rainfall.

Besides, it was feasible to create new agricultural lands through a plan of mountain-razing, mainly in areas of temperate or subtropical climate tending to be densely populated.

Mountain-razing, though costly, made possible an effective and permanent gain of flat land from useless crests as well as from the filling of gorges with the resulting debris.

Obstructed mountain streams forming ponds and pools at predetermined places were to increase fresh-water availability and soil humidification to feed underground streams and deposits.

Enlarging the arable land in rugged countries was a great challenge. It literally meant to make the planet's surface grow for significant purposes.

Successful survival required maturity
to realize that the earthly human habitat was shrinking and decaying under the abuses of civilization, and that it was well the time for mankind to mature. The call was inescapable: to reach maturity before the decay of the medium left no medium where to mature.

The challenge of the times to come
to the human kind was mainly to be alive and not extinct, given the hazards of environmental pollution and decay, radioactivity, drug addiction, sexual perversions, and anticonceptives endangering the genetic powers of the race.

Nevertheless, the racial youth of men's kind
endowed millions of individuals with great vitality and adaptability, so that survivors would produce even fitter offspring in a less crowded world.

Then, the challenge to the victors
in the biological battle for survival was to create the means for attaining a highly refined style of living.

High refinement in the material order
was to be achieved by concerted rates of population renewal to forestall hunger and unemployment, and by the balanced output of food, manufactures, services and energy, for unemployment and hunger were a shame to every human group and because energy crises could paralyze civilization precisely in times of generalized prosperity.

A high style of living
in a physically conditioned medium was to be truly high, essentially for its ethical standards.

Ideal norms of behavior
conveyed the notion that everyone was a unit belonging to a whole—a family, a nation, the human race—with all the rights to enjoy the wonder of living, and with all the responsibilities to be beneficial to society.

The one condition
for every ethical norm to be guiding and inspiring was veracity. The truth needed no interpretation. The truth only needed clarity and honesty to make men self-reliant.

The truth about the initiation,
the course, and the conclusion of vital processes, up to everyone's understanding, had to be told, according to the incontrovertible logic of observable facts.

Life on earth and elsewhere
had to be understood as the objective wonder it was, not as an oddity out of human comprehension.

The supreme ethical standard
held that everything promoting, improving, protecting, enhancing life was good. Everything destroying, degrading, perverting, exploiting life was evil.

In its substance and nerve,
life was objective. In its nothingness, death was subjective. Coming in peace, death was an endless sleep. Coming in torment or long suffering, a final release from pain.

The presence of life was tangible.
Carved from marble or cast in bronze, the likeness of human beauty was the symbol of life transcending time. Followed only by absence, in its thrill and thirst, in its mirth and sorrow, life was life but once.

As life was inevitable
and bound to grow and succeed, the meaning of the great challenges was plainly to rally courage to advance, true to the highest ethical standards and the farthest practical aims.

XV. – A New Cosmic Power

Three million years after emerging, mankind had a long way to go. Dinosaurs had dominated for a hundred million years, but were unable to withstand the environmental changes of the Cretaceous. Men could thrive on other planets, and even in space.

General decay or partial destruction of the medium could not have caused outright extinction of people. Biological adaptation to a degraded environment would have permitted survival, although in sorry conditions of irreversible degeneration.

On the total destruction of the land medium by nuclear warfare, what would have survived, once the seas still flowed, had to have been marine life, even at the high levels of radioactivity. Again, aquatic plants and animals would have evolved to enliven the desolate land through the power of life to overcome hostile physical conditions.

Universal natural cataclysm discarded as improbable, atomic war was a more possible way to disaster, but the powers capable of causing it showed no enthusiasm for suicide. By the turn of half a century of atomic power development by several nations, peace by dread seemed likely. A final weapon had proved more dissuasive than generations of preachers and pacifists.

Inevitable peace having befallen,
a sterile coexistence ensued. In the status quo, no one would move forward. As long as the inertia dragged on, civilization as a whole could make no great stride forward.

Mutual suspicion
among major and minor antagonistic powers led to armamentism and kept draining the world's economy, ruling out every chance to improve the quality of life except for those ensconced in official positions and big business, legal or illegal.

The atomic standstill, however,
was not to paralyze mankind for long. The world societies, mainly in the emerging countries and in those recovering their liberties, were plainly participating in the changes that turned out to rush without the help of a third world war.

The time of political regimes
forcing their subjects to accept the inacceptable was fast running out. Yet, it would have been too optimistic and passive to expect all the victims of injustice to gain freedom and well-being just by themselves.

If the free nations on earth
were plain to acknowledge the division of the world into one part free and well-off, and one part bond and hungry, then the human race would have to wait much longer for the universal community of free men bound to dawn sooner or later.

Individual liberty, though,
was not an end in itself, as it had been the natural condition of men. Rather, it was the love of liberty the passion that led men to the noblest achievements and most heroic deeds. Creators and discoverers dispelling the darkness of ignorance were men with a burning passion for liberty.

Also intensely human
was the instinct of association. Associated under a flag, a symbol, and a dream, men created splendid civilizations where power was based on effective civil liberties.

And the nation that was to be built
by all the peoples, throughout the fatherlands on earth, was a single mighty nation of free men.

In order to take the notion
of a world government into the field of realizations, a federate administrative system had to be projected in great detail prior to the fall of the last arbitrary political regimes in the world.

The pan-national government system
required to be no political novelty. It rather amounted to a pragmatic synthesis, applied on a widely enlarged scale, of what human societies through the centuries had devised as the fairest methods of civil administration, from the tribal stage to that of representative federated republics.

The consolidation of continental unions
bound to incorporate in a single pancontinental federal republic was thoroughly feasible once the two most powerful unions federated.

The possession of real economic power
and the control of key geographical zones were to establish the unions already incorporated in a position leaving the remaining nations no choice but to incorporate, for under no other system could their true interests be better served.

Considering that the distances
formerly separating the lands of men had kept nations apart and differentiated, and that the fast progress of communications had shortened those distances to absolute minimums, uniforming life-styles, needs, and aspirations, it was sensible to foresee that the world moved irresistibly toward universal political unification—each nation, of course, keeping alive her traditions, her glories, and her pride.

But although events in the age of electronics
proceeded faster than ever, general unification processes were to take much longer than many decades.

The fact that spirits of selection
had for long been aware of the need of world unification and even ready to participate in the great changes did not mean that most people had a notion of the developments lying ahead.

Poverty and ignorance kept much of mankind
in a lagging condition that was to persist if effective action was not taken to turn hopelessness into real expectations.

The promotion of the human species
was the concern of men and women of vanguard acting from every position of influence, as parents, educators, authors, philosophers, statesmen. With them, acting concertedly, people everywhere were to advance vigorously. Without them, the race would have lost the title of human long ago.

With ever greater force,
the minds of vanguard were to upgrade common people of all races until making them aware of their worth as members of a vital species, fit for attaining very high objectives.

Likewise, it was greatly a matter
of imbuing everybody with the notion that once federated, each nation of the world would grow richer, not yet because of the enhanced ability of a united world to produce riches and create well-being, but because of the colossal savings to pile up from suppressing diplomacy, armies, arms, wars.

Quite understandably,
the actual declaration of unification of the states of the world was to take much shorter or much longer, subject to vastly complex circumstances and extremely powerful forces for and against, certainly more against than for.

The more complex a society grew,
the more profoundly circumstances shaped its development. Most great leaders could rise to power and command climactic events because special circumstances made their talent or their courage good at precise moments of history.

Furthermore,
circumstances might be accumulative as well as accidental. Accumulative circumstances were the intensified traveling through the mechanized era, and the contacts and relationships developed among persons from every land of the earth.

The epochal consequence
of the intense interchange of ideas was to be nothing less than a generalized notion of the community of feelings and expectations of humans as a whole.

Accidental circumstances
could have been world crises and starvation; new kinds of epidemics; the emergence of mysterious prophets spreading fanaticism for weird causes; total war; natural disasters like global drought or flood, a sudden glaciation, or an eventual heating up of the world.

Powerful forces driving mankind
toward its great destiny were accumulative as well as evolutionary. The diffusion of objective knowledge led men to an unstoppable process of universal enlightening, which was intensely accumulative and dynamically evolutionary.

Evolutionary were the needs of communication, cooperation, interchange between close neighbors and among distant associates in common interests.

Strongly negative
since the formation of nations were the forces of conservatism, zealotry, vested interests. The forces of regression had for long been in charge of repressing tendencies to reform and of persecuting heretics and innovators.

Still, it would have been practically impossible
that the negative forces had always won. Deep in the human nature was an instinct leading to liberty and advancement to be manifested in ever new ways through the ablest and the bravest individuals in every race and every nation.

The fact that the species
had unquestionably advanced from pre-human conditions to highly complex forms of civilization suggested that, after attaining full physical control of the medium and the political unification of the world, men would advance more than ever, eventually toward fulfilling their natural role as a new cosmic power.

Because a time had to come
when universal education, cultural advancement, science, and the technical arts, having solved every pressing problem of life on earth, vitally needed wider spheres of action.

No longer spoiled in wars
or minimized by hunger, disease, or ignorance, the titanic energies of a unified human race were to find fitting outlets in the conquest of other worlds and in the human projection into the depths of the times to come.

The character of transterrestrial enterprises
was compatible with the scientific and financial might that a planetary power was to be able to build up for the achievement of supreme objectives.

The fascination exerted by distant worlds
had motivated intense astronomical observation through the centuries. As a result, increasing amounts of information were available on the nature of every member of the solar family.

Thus it was known that the giant planets,
from Jupiter to Neptune, being too distant from the sun, were cold and fluid, as their gaseous elements could not evaporate. The giants so amounted to enormous deposits of hydrogen, methane, ammonia.

The small planets
circled the sun more closely and were terrestrial, since they presented solid surfaces. Most of their gaseous matter was lost by Mercury, the Moon, and Mars, but was retained by the somewhat larger masses of Venus and Earth.

On the purpose of establishing human life
beyond the earth, no sister planet was ready to be inhabited. Closest to the sun, barely observable, and rotating very slowly, Mercury was little larger than the Moon.

The awfully hot, long Mercurian days
alternated with the coldest nights in the Sun's dominions, letting note that stellar energy traveled through space as radiation, not as heat.

Venus also rotated slowly
and was shrouded in a thick atmosphere resulting not so much from primary gaseous components, but from carbonates being gasified under the strong solar radiation. In the dim Venusian inferno, there were no days and no nights.

Revolving next to Venus,
the Earth was laden with water and life, but its dull Moon was as barren as any other dry world. Virtually forming a binary system with the Earth, the Moon had to be the first human outpost in space for all its forbidding characteristics.

Of the solid planets
escorting the Sun in inner orbits, Mars was the coldest, as it was the farthest. Prolonged dust storms eroded the dry surface and were covering Mars with roving dust layers. Large underground water deposits produced rains every hundred thousand years, but were no help to the development of Martian life, as photosynthesis would not take place underground.

Pluto was the loneliest body in the Solar System. As an outer planet, Pluto was small, and it had probably been a satellite of a giant planet.

The small bodies orbiting most planets, same as those orbiting the sun, like the asteroids, were rather chunks of rocky material, many of them resulting from collisions between minor bodies when their orbits were still undefined. Only major satellites were spherical.

The prospective worlds for human expansion were initially reduced to Mercury, the Moon, and Mars, although Mars's share of sunlight was to be poorer as the solar radiance decreased.

The possession of the three neighboring worlds was to open unimaginable new fields of action and untouched mineral resources. Controlling the Moon was thus the first major commitment in the operations aimed to put new worlds within the reach of the insatiable human spirit of adventure and enterprise.

The difficulties involved in the operations of interplanetary expansion were to be remotely greater than those met in previous epochal conquests, as all had been undertaken within the cozy earthly medium, except the U.S. astronauts' lunar promenades and the Soviet crews' long stays in stations orbiting the Earth.

The Moon, Mars, and Mercury
had no air, no surface water, not even enough gravitational pull for a normal physical course of earthly life.

So, while plant life had all by itself
made the Earth livable through the processing of water, carbon dioxide, and sunlight, men had to make barren worlds fit for life only in some cases counting on native underground water deposits. On the long lunar days and on those even longer of Mercury, however, solar radiation was more than abundant.

Distances separating the inner planets
were shorter than those among the outer planets and could be covered in months at the conventional speeds of spacecraft traveling by gravitational impulse as solar satellites.

To shorten interplanetary distances
it was inexcusable to contrive faster spacecraft, which was equivalent to the advance from sailboats to motorboats.

In fact, by short impulses of nuclear propulsion,
space-travel speeds were to be increased, then kept up by the principle of inertia, there being no resistance in the void—up to very high rates of motion.

Still faster space ships
were subsequently to take expeditions to neighboring solar systems at speeds in a measure nearing that of light.

In an age of intense space travel,
take-offs from the Moon were to be less costly, for the escape velocity from the Moon was about a fifth of that from the Earth.

Given its value as the Earth's spaceport
and first cosmic possession, the lunar world had to be adapted to sustain terrestrial life and to allow the execution of large-scale operations leading to establish a working binary system, the Moon thus becoming an essential complementary world to the expanding earthly civilization.

The very first step
toward making earthly life possible on the Moon was to provide for the vital needs of shelter, oxygen, water, and food—all in a world apparently lacking every life-sustaining condition.

To start with, however,
the lunar surface received sunlight from the same moderate distance as the Earth did. Water also occurred on the Moon, as it was universally produced by hydrogen and oxygen occluded in rocks and other elements that happened to undergo high temperatures.

As a sphere of igneous formation
with a fourth of the Earth's diameter, during its brief geological youth, the Moon produced large amounts of water, which filled underground deposits and extensive shallow seas.

Its gravitational force being too weak
to hold any atmosphere, the Moon soon lost its seas by evaporation, but the empty seabeds remained as the dark lunar maria, attesting the Moon's brief marine past.

The strong earthly gravitational influence
kept the Moon exposing only its heaviest side to the Earth right since its capture as a young minor planet that had been circling the Sun too close to the Earth's orbit.

Obviously, the lunar underground water deposits avoided evaporation and remained, their presence being detectable by the myriads of rilles, mainly throughout the maria, indicating that water occasionally filtered out into the dry surface only to flow briefly and evaporate.

Determining whether the Moon
was still hot inside and producing any new water, and how much water remained underground was to be critical to the cost of the Moon's conquest.

The rather improbable eventuality
of there being no water left on the Moon would imply a higher cost of the lunar operations, since water had to be produced from materials containing hydrogen and oxygen, all abounding in the composition of igneously formed spheres, and undoubtedly occurring on the Moon.

Apart from mineral riches
for eons awaiting utilization, the Moon had a great asset in the enormous amounts of solar radiation striking its surface through the 327 hours of the lunar day. So, all the energy required by human activities on the Moon was to be satisfied with electric power generated by solar batteries.

With oxygen, water, and energy
definitely available on the Moon, vital living space with ideal constant temperatures both for residence and farming was to be found in underground galleries left by old water deposits.

Life in the pressurized, climatized galleries
was to be largely self-sufficient, with sunlight let in by wide filters screening out lethal radiation, abundant electric power for the long lunar nights, plants producing food and oxygen, while carbon dioxide given off by lung respiration and other kinds of combustion was to feed plant life, plants themselves descending from the most adaptable and productive earthly species.

Later, cities not too different
from those back on Earth were to dot the lunar landscape, interconnected by a network of highways, so that Moon residents could easily shift from day areas to those in the night or twilights. Meanwhile, underground farms yielding ever more food and oxygen were to multiply, as more underground space was discovered or simply excavated.

Of course, humans stationed on the Moon were not to live only in enclosed quarters, for life outdoors was to be intensely dynamic during the eighty-hour long early morning and also during the equally long early evening, when there was enough sunlight, yet the Sun did not strike hard.

In light spacewear, life on the Moon was to be full of action, both for business, making the Moon hospitable and attractive, and for fun and sport, as the gentle gravitation allowed very high jumping and fast running. Individual flight by knapsack rockets was to invite to adventure and also to facilitate the exploration of every tract of lunar land for highly detailed mapping.

Undoubtedly, with the busy spaceports receiving and dispatching ships from and to the Earth and the worlds to be conquered next; with the active mines, factories, and farms exporting minerals, manufactures, and extraordinarily sized farm products, the conquest of the Moon was to imbue the human race with a new sense of realization, the Moon thus having performed its role in a unique binary system, bound to extend its power over the central planets of the Solar System.

The experience of having submitted the Moon to the human power was to be invaluable for the planning of successful operations on Mercury, then on Mars and beyond, although under widely different physical conditions.

As a foremost observatory
from which to monitor the Sun's activity, and as a close receiver of highly concentrated solar radiation, Mercury well deserved to be conquered as men's second cosmic possession.

Operations on Mercury were to be initiated
from underground bases located in polar areas, where the strong solar rays struck very obliquely, there being not much axial tilt, as Mercury rotated very slowly. Likewise, exploratory missions had to proceed toward lower latitudes along the slow-advancing dawn twilight belt, and only while the slant torrent of sunlight dawned over the cold surface emerging from the 708-hour long Mercurian nights.

Later, more numerous missions to Mercury
were to occupy lower latitudes, lodged in galleries arranged after the lunar models, but located at lower levels for protection from the extreme temperatures. Life outdoors thus had to take place only at twilight hours of the prolonged Mercurian day. Water and oxygen could only proceed from sources and methods largely similar to those already used on the Moon, as Mercury had also an igneous geological past.

The wide thermal differences
between the long Mercurian days and nights made the planet amazingly rich in solar energy, which was convertible into all the amounts of electric power to be needed in the profitable conquest of the singularly challenging Mercurian world.

Orbiting the sun as the farthest inner planet, Mars, with only the asteroid belt between, was next to Jupiter, the huge gaseous ball that could have been a star.

The Martian world was cold, dry, and windy. However, there existed all the physical elements to sustain life. Under the dusty surface lay vast deposits of liquid water and several classes of hydrocarbons, once Mars, as all the inner planets, had an igneous geological history.

The Martian gravity, being half as strong as Earth's, could hold only carbon dioxide, a heavy gas that was good to feed plants and yield oxygen upon dissociation from carbon. The gas was also good to retain heat, and certainly helped make the Martian climate less brutally cold.

For protection from the low surface temperatures and dust storms, living quarters for the Earth's missions appointed to Mars had to be established at certain depths under the surface, where the planet's internal heat provided stable warmth. For the rest, the knowledge acquired while submitting the Moon and Mercury was to make of the survival in strange worlds a fast-advancing specialized science.

The energy for producing everything necessary to consolidate the possession of Planet Mars was to come from putting its hard-blowing winds to drive electric generators, as the heavy carbon-dioxide atmosphere conveyed large amounts of kinetic energy.

After long periods of human permanence
in other worlds, it had to be reasonable to expect biological processes of adaptation tending to produce modified types of men. Periodic returns to Mother Earth were thus to be prescribed on the vital purpose of keeping the genetic unity of the human race.

The magnitude of the human conquests
in all the fields of knowledge and in those of practical undertakings had to be revealing of the forceful human temper, which would not accept the eventuality of biological extinction so unknowingly and fatally as dinosaurs did.

Far from that,
men united in a single mighty nation were to be capable of improving everything, chiefly their own genetic standards and the environmental quality of their native world.

The conditioning and beautifying
of nearby barren worlds for human residence, as well as for bases of operations into the wonder of the spheres shining beyond the Sun, were to be genuine expressions of the rise of a new kind of cosmic power acting on its own designs, in true accordance with the physical laws of the universe.